LIVING WITH HERDS

Human–Animal Coexistence in Mongolia

Domestic animals have lived with humans for thousands of years and remain essential within the everyday lives of people throughout the world. In this book, Natasha Fijn examines the process of animal domestication in a study that blends biological and social anthropology, ethology, and ethnography. She examines the social behavior of humans and animals in a contemporary Mongolian herding society. While living with Mongolian herding families and their herd animals, Dr. Fijn observed both sides of the human–animal relationship. Examining their reciprocal social behavior and communication with one another, she demonstrates how herd animals influence Mongolian herders and how the animals themselves are active partners in the domestication process.

Natasha Fijn is a researcher and observational filmmaker. Her main research focus is on human–animal interaction and social engagement. She currently lectures in Visual Culture Research at the Australian National University. Natasha has a background in anthropology, zoology, and wildlife filmmaking. She has worked with the BBC, Natural History New Zealand, and Green Umbrella Productions on wildlife documentaries and is now working as an independent filmmaker.

LIVING WITH HERDS

Human–Animal Coexistence in Mongolia

NATASHA FIJN

Australian National University

CAMBRIDGE
UNIVERSITY PRESS

CAMBRIDGE
UNIVERSITY PRESS

University Printing House, Cambridge CB2 8BS, United Kingdom

One Liberty Plaza, 20th Floor, New York, NY 10006, USA

477 Williamstown Road, Port Melbourne, VIC 3207, Australia

314-321, 3rd Floor, Plot 3, Splendor Forum, Jasola District Centre, New Delhi - 110025, India

79 Anson Road, #06-04/06, Singapore 079906

Cambridge University Press is part of the University of Cambridge.

It furthers the University's mission by disseminating knowledge in the pursuit of education, learning and research at the highest international levels of excellence.

www.cambridge.org
Information on this title: www.cambridge.org/9781108431057

First published 2011

A catalogue record for this publication is available from the British Library

Library of Congress Cataloging in Publication data
Fijn, Natasha, 1975–
Living with herds: human–animal coexistence in Mongolia / Natasha Fijn.
p. cm.
Includes bibliographical references and index.
ISBN 978-1-107-00090-2 (hardback)
1. Ethnology – Mongolia. 2. Herding – Mongolia. 3. Domestication – Mongolia.
4. Human–animal relationships – Mongolia. 5. Mongolia – Social life and customs.
I. Title.
GN635.M66F56 2011
306.09517′3–dc22 2010038922

ISBN 978-1-107-00090-2 Hardback
ISBN 978-1-108-43105-7 Paperback

Additional resources for this publication at www.cambridge.org/9781108431057

CONTENTS

PART THREE: LIVING WITH HERDS

LISTS OF PLATES, FIGURES, MAPS, TABLES, AND ONLINE VIDEO SEGMENTS

PLATES

FIGURES

MAPS

TABLES

Appendix Tables

ONLINE VIDEO SEGMENTS

www.cambridge.org/9781107000902

GLOSSARY

i. General Mongolian terms

Ail	one family's home
Am	mouth
Am'd	anything that lives or is animate
Am'tan	animal (often wild)
Arat	folk method of breeding livestock
Arkhangai	Mongolian province within the Khangai; an area of one of the field sites
Baigal	nature
Baigal' gazar	landscape or place
Baigal' orchin	environment
Bulgan	Mongolian province within the Khangai; an area of one of the field sites
Deel	a long garment, similar to a coat or dress
Em dom	magical medicinal cure; medicinal home remedy
Ger	home, that is, a nomadic tent, equivalent to the yurt
Geriin tejeever	to feed or nurture within the domestic sphere
Im	earmark or symbol
Jakhdag	a lock of hair
Khadag	blue, silk sash
Khangai	Mongolia's central mountainous region
Khirigsuur	ancient stone monument and grave site of Turkic origin
Khot ail	family encampment, domestic sphere
Khusuur	scraper, made from wood or pelican beak; used for scraping sweat from horses or as a brush for dislodging dirt from a horse
Mal	domestic animal or herd animal
Malchin	herder
Maliin banzuragch	God who protects the five kinds of animals

Mal süreg	to herd
Morin Khuur	Mongolian horse-head fiddle
Naadam	festival or celebration
Negdel	a cooperative or a communal farm during the Soviet era
Otor	a temporary camp where herd animals graze on better pastures
Övöö	sacred rock cairn on prominent mountain top
Saalchin	milker
Seter	Sacred or holy animal (Tibetan, *tsehtar*)
Shilbuur	elongated whip
Soijakh	hardening animals for winter
Tamga	brand or stamp on horse's hindquarters and the branding iron itself
Tavan khoshuu mal	the five kinds of animal
Tengger	the sky, the heavens, a god, the weather
Tsagaan Sar	White Month, or Mongolian New Year celebration
Tsetserleg	garden; central town in Arkhangai Province
Udgan	a woman shaman or healer
Uurga	horse lasso-pole
Zud	severe weather during winter
Züs	external features; coat colour

ii. Mongolian dairy products[1]

Aarts	curds from sour milk, made from boiled yoghurt with a portion of milk: The curds are placed in cotton cloth and pressed between boards to remove the whey. May be stored in sheep stomach and used during the winter.
Aaruul	curds that are pressed, sliced, and then dried, with or without sugar added: They are dried on a board in the open air on the roof of the *ger*, or hung inside above the fireplace. Yak *aaruul* is made into larger portions and contains a higher fat content.
Airag	fermented mare's milk, known as "white beer" or "koumiss": Fresh mare's milk is poured into a cowskin bag. Milk is added to an initial culture throughout the day and then churned 800–1,000 times and left to ferment overnight. *Airag* has a sweet–sour taste and fragrant smell.
Byaslag	cheese made from warm milk with a small portion of yoghurt: After the milk turns to curd, the mixture is placed

[1] For more detailed information regarding some of these dairy products, see Indra (2003).

	in cotton cloth and pressed between boards to remove the whey (the cheese is not salted or aged the way many cheeses are).
Eezgii	yoghurt added to milk to become sour, to curdle: *Eezgii* is the residue on the bottom after boiling.
Nermel	distilled spirit from fermented milk: The fermented milk is placed in a wooden barrel and repeatedly heated to distil a liquor from the condensation that forms inside the barrel. The liquor has an alcohol content of approximately 10–12 percent.
Öröm	thick cream: Milk is heated and aerated with a ladle and left to congeal for twelve to twenty-four hours; then the cream is scooped off the top. The substance is often eaten with bread, tea, and deep-fried dough for breakfast.
Shar tos	"yellow oil," or boiled butter: It is made from fermented cream that has been stored in sheep stomach or wooden tubs.
Suutei tsai	brick tea made with milk and salt: The mixture is poured from a height until the milk has homogenised.
Tarag	yoghurt: It is made from a portion of previous yoghurt culture added to warm milk, covered with cloth overnight to activate the culture. Obtained from cow, goat, and sheep milk in summer.
Tsagaan tos	"white oil," derived from fermented cream: Boiled butter is mixed with *eezgii* and flour, hot tea, or water. It is often eaten with sugar.
Tseghe	milk-based alcoholic drink similar to *airag* but made from cow, sheep, or goat milk: Milk or water is often added to the drink to reduce sourness.
Tsotgii	unboiled milk made into cream: It is made from cow, sheep, and goat milk.
Tsotgiin tos	butter: It is made from *tsotgii*.
Uurag	colostrum, or rich protein milk after the birth of young: It solidifies into a creamy, jelly-like consistency, slightly sweet and similar to egg in taste.

iii. Key animal terms[2]

Aduu	Mongolian horse
Argali	mountain sheep

[2] For a dictionary of Mongolian animal husbandry terms, refer to Dorjgotov (1999).

Azarga	stallion, uncastrated male horse (also a horse herd, including a stallion with mares and young)
Bod	horses, camels, and "large cattle," such as yak
Bog	sheep and goats
Bukh	bull, or uncastrated male horse
Byaruu	two-year-old cattle
Daaga	two-year-old horse
Em khon'	ewe, female sheep
Em yamaa	nanny, or female goat
Er byaruu	bullock, or young male cattle
Gunj	heifer, or a young cow that has not yet calved
Güü	mare, or female horse
Ireg	wether, or castrated male sheep
Khainag	cross-breed between a yak and a cow
Khuts	ram, or male sheep
Mor'	gelding, or castrated male horse
Ortoom	second-generation cross between yak and cow
Sarlag	yak, including both male and females
Serkh	wether, or castrated male goat
Shar	ox, or castrated male bovine (yak, yak–cow hybrid, Mongolian cattle)
Takhi	Asiatic wild horse – known as Przewalski's horse
Temee	Bactrian camel
Ugjikh	bottle-fed lamb or kid
Ukhna	buck (or billy), or a male goat
Yangir	mountain goat

iv. Key behavioural terms

Aashtai	moody, bad tempered, ill bred, momentarily angry
Agsam mor'	a horse with mettle, or spirited and lively
Darkhlakh	protector of the herd
Dogshin	impolite, difficult to control, perhaps violent, fearful, unbroken horse (opposite of *nomkhon*)
Dölgöön	calm, peaceful (*dölgöön aash* is an animal with a gentle disposition; *Dölgöön* is also a common name)
Emneg	untamed, similar to *khangal* (often applied to oxen)
Erkh	self-willed, pampered (but can also mean "tame" with animals, or "spoilt" in relation to children)
Khangal	wild, untamed, untrained, unbroken horse
Khatuu amtai	hard-mouthed horse, or a horse that pulls on the reins and is unresponsive

Khatuu khamtai	hard-nosed ox, unresponsive to signals
Khenz	individual animal that is small and cute, literally meaning "late born"
Khöngön khöltei	easy riding horse, or an ox that is light footed and quick stepping
Muu	bad, not good, useless, evil (or *muukhai* when the animal butts and kicks)
Nomkhon	tame, quiet, calm or placid (key word to indicate a good animal; the antonym of *aashtai*)
Omgolon	restive, impulsive, difficult, arrogant, or haughty
Oroo	shy, elusive (often in relation to horses, *oroo dogshin* means spirited whereas *oroo khuljaa* means a runaway horse)
Sain	good, fine, kind, or nice
Sergelin	quick witted, alert, or astute (*sergelin galzuu* can mean "a bit crazy")
Shildeg	selected, the best; used in relation to racehorses
Shiruun	tough, rude, rough, violent behaviour (similar to *omgolon* and *aashtai*)
Shudenge mor'	green, or newly broken-in horse (also referred to as *sain nomkhroo ügüi mor*)
Taivan	calm, gentle, peaceful (similar to *dölgöön*)
Tolgoilogch	leader
Tsogio mor'	horse that is a galloper
Zalkhuu	lazy, or idle
Zerleg	wild, or not habituated to humans (can also mean "crazy")
Züülön amtai	soft-mouthed horse, quiet when ridden (similarly referred to as *am zöölön surguul'tai*, as a manageable, or trained horse)
Zöölön khamtai	soft-nosed ox, quiet and easy to pull along or hitch behind a cart
Zörüüd	stubborn or obstinate (often a loner)

ACKNOWLEDGMENTS

I will begin by thanking the first Mongolian family I met, Nyamdorj's family, who resided in Canberra, Australia. Delgermaa gave me lessons in Mongolian and translated for me, even though she was often busy. I am deeply indebted to many people in Mongolia for their willingness to accept me into their lives and for their unbounded hospitality. Thanks to Tuul for taking me in when I needed accommodation in Ulaanbaatar and involving me with all of her family. She and her family were always welcoming when I arrived at short notice from one of my field sites, or even after I had been absent for more than a year.

There are many people that showed me great kindness and helped make my life easier in Mongolia, even in small ways. It is due to this unbridled hospitality that I had a remarkably untroubled field experience, as I felt safe and never felt ill at ease while living in Choijo's or Dogsomjav's encampments. I could not have written this book without the daily communication and engagement I experienced with the herd animals while I was in Mongolia, whom I came to know as individuals and as real characters and personalities.

Thanks to Choijo's extended family for their acceptance of me into their lives, for allowing me to film, to ask endless questions, participate in herding activities, ride their valuable horses, and live in their *ger*. Here, I would particularly like to extend my gratitude to Naraa, who accepted me as a daughter. I am grateful to Dogsomjav's extended family for being so accepting of my presence in their lives. Dogsomjav and Lhagva were particularly helpful with all aspects of my research and in making me feel at home in their encampment. I value that Dogsomjav often still speaks of how I used to carry her full pails after milking and how I was good with the cows. She is inspirational in her strength of character and as a tireless worker. Thanks to her daughter Davaa for working as my field assistant during the summer of 2005, for always making me feel welcome in her home in Erdenet, and for keeping me connected to the rest of her family.

Professor Tumen and Erdene from the Mongolian State Agricultural University assisted in obtaining a research visa and initially connecting me with Choijo's extended family. I am indebted to Enkhtuvshin, Altansukh, and Divangar from the Mongolian State Agricultural University for assisting with finding a second field location, accompanying me into the field to meet herding families, providing me with valuable information for my research, and for their friendship while I was in Ulaanbaatar. Thank you to Baigalaa and Uka, who worked so hard to assist and interpret in the field. Urnaa and Erdenekhuyag helped me to visit herding encampments and to observe horse training and racing events during the winter of 2005.

This book is based on an original manuscript that was submitted in June 2008, as part of a Ph.D. thesis in anthropology at the Australian National University. I could not have asked for better advisors on this project and ultimately in the completion of the manuscript. Beatrice Rehl from Cambridge University Press was immediately interested in my book proposal. People at the Australian National University who gave valuable advice along the way included Robert Attenborough, David MacDougall, Judith MacDougall, Nic Peterson, Gary Kildea, and Pip Deveson. Thanks particularly to Colin Groves and Deborah Rose for being so supportive and providing invaluable feedback. Breakfasts with the Eco-humanities Group provided many inspiring and enlightening discussions. Thank you to the great staff within the School of Archaeology and Anthropology at the Australian National University. The university provided me with an Australian Postgraduate Research scholarship, a Field Work Grant, and a Travel Research Grant, which made this ambitious project possible. Fraser Argue, in association with ActewAGL, sponsored and assisted me in setting up a portable solar panel unit, which enabled me to recharge batteries and run a laptop during my time in the field in Mongolia. Shawn Laffan created the map of Mongolia at the beginning of this book.

I would not have enjoyed the experience of writing this book nearly as much as I did without the support of friends within anthropology and related disciplines at the Australian National University, particularly Hannah Bulloch, Angie Bexley, Debbie Argue, Justin Barker, Josh Wodak, Sarinda Singh, Mike Fabinyi, John Carty, Silke Andris, Greg Dvorak, Thom van Dooren, Ma Khin Mar Mar Kyi, and Jinghong Zhang. Ximena Nelson who read parts of the manuscript and provided valuable ethological and methodological advice. Kerry Weston went to considerable effort to locate and to actually visit me while I was in the Mongolian countryside.

My mother Anne Gillespie and my father John Fijn have always been extremely supportive. I value that Dad encouraged me to embark on this project in the first place and that Mum provided constant emotional support, especially while I was living in Mongolia and was hard to track down. A big thanks to my partner, Julian Laffan, who has lived through the experience of this project with me from start to end. Orlando Louis Fijn Laffan literally lived through every moment during the last few months of writing as a little growing being in my belly. Last but not least, Billy (the dog) provided constant observational entertainment and inspiration while I wrote this book. This book is dedicated to the animals who, as individuals and sentient beings, have enriched and shared many aspects of my life with me.

Author's Note

I took all photographs and digital video footage, which supplement this book, during my time in the field in Mongolia (during 2005 and the spring of 2007). Throughout this book Mongolian words are designated in parentheses and italicised. I have referred to the names of Mongolian protagonists by the shortened versions of their names that were used in an everyday sense, unless informants asked to be referred to by their full name within this book. The translation of Mongolian words is written in accordance with the grammar of *Colloquial Mongolian* (Sanders & Bat-Ireedüi, 1999), such as the use of "kh" instead of "h" and elongated vowels instead of the shortened form, such as Ulaanbaatar instead of Ulan bator. Mongolian has "front" and "back" vowels. The front vowel of "ö" is pronounced as the "yea" in "yearn" and "ü" as the "u" in "put." I have opted to designate the "shortened " sound at the end of some words as a single quotation mark, as in "*mor'*" (the word for gelding).

LIVING WITH HERDS

SPRING

Plate 1. Brewing spring storm.

Plate 2. Feeding milk to a lamb with an ox horn.

Plate 3. Uuganaa with new-born calf.

Plate 4. Neighbouring herders combing cashmere from a goat.

Plate 5. Saikhanaa returning from herding.

Plate 6. Ewes and nannies nursing their young within an encampment.

Plate 7. Khorlo handling a two-year-old (*byaruu*).

Plate 8. Naraa milking a *sarlag* by hand.

PART ONE

CROSSING BOUNDARIES

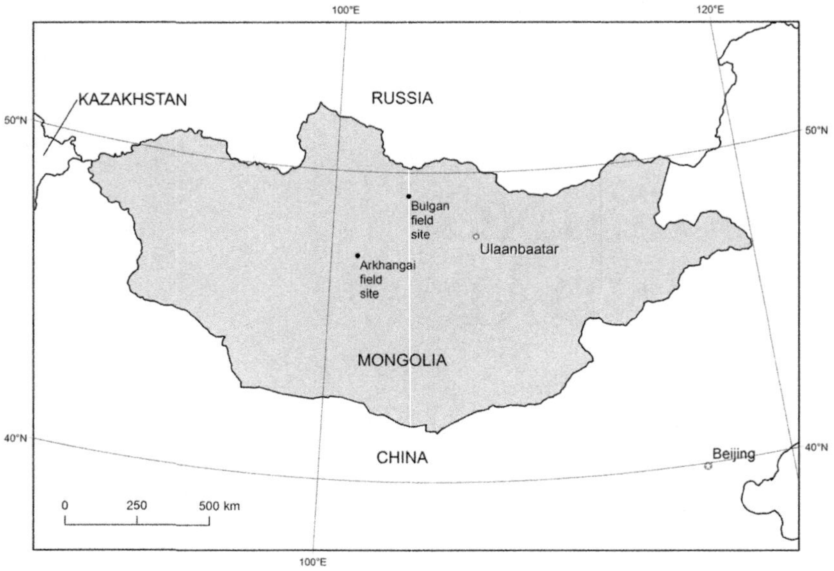

Map P.1. Mongolia, including the location of Arkhangai (west) and Bulgan (north-west) field sites in relation to the capital of Ulaanbaatar. (Source: Created by Dr. Shawn Laffan, University of New South Wales, 2010.)

PROLOGUE

LIFE IN THE KHANGAI MOUNTAINS

Farming and city-building peoples have always looked down on nomads as "primitive"...but this does not mean that nomads have reciprocated by looking up to the city and farm. Far from it. There is a strong tradition among nomads that they are people who have made themselves free from the drudgery of the peasant and the shut-in life of the city dweller. (Lattimore 1962: 34)

FIRST IMPRESSIONS OF MONGOLIA

I could see the Soviet influence on the buildings in Mongolia, as a former satellite state of the Soviet Union. The Mongolian capital's expansive main square was flanked by parliament on one side and a drama theatre with an imposing façade and huge concrete columns on another. Every drab apartment block had the same basic design with thick double doors and double-glazed windows. This environment was very different from my expectations of a Mongolia with no locks on brightly painted *ger* doors and sparse, expansive grasslands.

My first few weeks in Mongolia in November 2004 were isolating and claustrophobic. The contrast between the below-freezing temperatures outside and the intense central heating inside gave me a feeling of suffocation. I suffered through episodes of feverish sweating from a terrible flu that developed soon after my arrival and was sure my fever was heightened by the fluctuating temperature changes. Outdoors, the temperature was often below -10° C and the streets were treacherous with slick black ice on the roads and footpaths.[1] The cold would bite into any exposed skin on my cheeks and the moisture from my breath would form icicles on my eyelashes. The sky was a brilliant blue every day because it was too cold and dry to snow and the

[1] Temperatures start to drop below zero in October, sinking to -30° C in January and February, remaining below freezing until April (Mayhew 2001).

weak sun hung low in the sky, vanishing below the nearby mountain range in the afternoon so that darkness descended by 5 P.M. I would only briefly venture outdoors to buy food and found the nearby mini-supermarket had sparse pickings of pickled vegetables, frozen meat, and sweets. My breakfast consisted of semolina, which was sold as baby food. When I attempted to organise any aspect of my research, I was faced with a cold reception of stern faces and what I perceived as closed-minded bureaucracy that I suspected was an effect of the previous oppression of the Soviet era (1921–1990).

My impressions of Mongolia changed as soon as I moved into Tuul's household. I had just taken a two-week "survival course," an initial crash course in practical Mongolian, prior to a more in-depth two-month course at a private language college.[2] Tuul was one of my teachers, and when I inquired about possible Mongolian families to reside with during the winter months, she offered a room in her small apartment. Tuul was a single mother and her world revolved around her one-year-old daughter Mongoljin (meaning Mongolian queen). To me Tuul represents an urban Mongolian, breaking with the role expected of Mongolian women. At thirty-three she was a single parent studying toward a master's degree in Mongolian linguistics and she had supportive friends who were also single, independent, and career minded.

Tuul and her parents have always resided in the Mongolian capital and live an urban lifestyle, yet the spatial structure of Tuul's family was the urban equivalent of an extended family encampment. Within the same apartment complex, Tuul's mother, father, and second brother lived in one apartment, and her eldest brother and his family lived in another; they lived only approximately 200 metres from Tuul's apartment. This structure also incorporated semi-nomadism, for in the summer season the extended family would vacate the apartment complex and move out to their home on the outskirts of the city. They would then retreat back into the warmth of the centrally heated apartments once the temperature dropped below zero. Some herding families also adopted this strategy: They would move to a township (*süm*) during the colder months of the year, and then into the countryside for the short summer season. Some family members remain in the countryside all year to tend to the animals, whereas those who move into the town centre are often elderly or school children.

During the winter months, Tuul's eight-year-old niece lived with her to attend school in the city, and a teenage cousin came to stay over the

2 Mongolian is related to the Altaic family of languages and is closely related to Turkic languages. Mongolian has no relation to Chinese or Russian, although some Russian words have been adopted by Mongolians.

weekends.[3] Both of the young relatives helped with babysitting or housework in exchange for residing in the apartment. They would share a mattress that also functioned as seating for watching television, just as a bed in a *ger* is used as seating during the day. The central fire functions as a focal point within a *ger*. In the apartment the television, with its flickering light and low-level background noise, became the centre point around which the family would eat and socialise. The television was not watched with avid attention but tended to be on in the background and glanced at occasionally, just as a fire crackles and needs to be fuelled periodically.

In the next two sections I describe my initial impressions of my field sites, which were both situated in the Khangai mountain chain in central Mongolia in two separate locations: Bulgan Province and Arkhangai Province (*aimag*). Although the Bulgan and Arkhangai field locations are located less than 300 kilometres from one another, to travel between the two required a detour via Ulaanbaatar (refer to Map P.1 for a map of Mongolia).[4] Apart from the direct arterial routes leading from the capital, roads consist of multiple two-wheeled tracks in the grassland. The roads are representative of the infrastructure of the country, where Ulaanbaatar features all that is urban. The roads are initially sealed but quickly dwindle to dirt, then just tracks in the vast grassland steppe. Beyond the capital city, facilities in the local towns are dilapidated shells of another era.

BULGAN FIELD SITE: DOGSOMJAV'S ENCAMPMENT

I had been in Ulaanbaatar for the previous three months during the depths of winter, when temperatures sink to −30° C in the coldest capital city in the world (Mayhew 2001). By March 2005 the weather was gradually becoming warmer but would remain below freezing until April. I was anxious to get into the countryside to establish the locations for my research and meet herding families with whom I could potentially live over the coming months.

During the initial three months establishing myself in Mongolia, I met three young lecturers at the Mongolian State Agricultural University who

[3] As the Mongolian school system is based on the Soviet education system, children attend school in urban centres for nine months with a long summer break from June through August. The state provides education free of charge to all Mongolians. According to UNESCO (Yembu and Munkh-Erdene 2005), 97.8 percent of Mongolians were literate in 2005.

[4] The distance between Ulaanbaatar and the Arkhangai field site was approximately 480 kilometres to the west, whereas to travel to the Bulgan field site was approximately 420 kilometres to the north-west. This meant that to travel between the two sites required four days of transport over rough roads (the Arkhangai field site was located at 47° 16′ 29″N, 101° 14′ 36″E; the Bulgan field site was located at 49° 15′ 11″N, 103° 37′ 23″E).

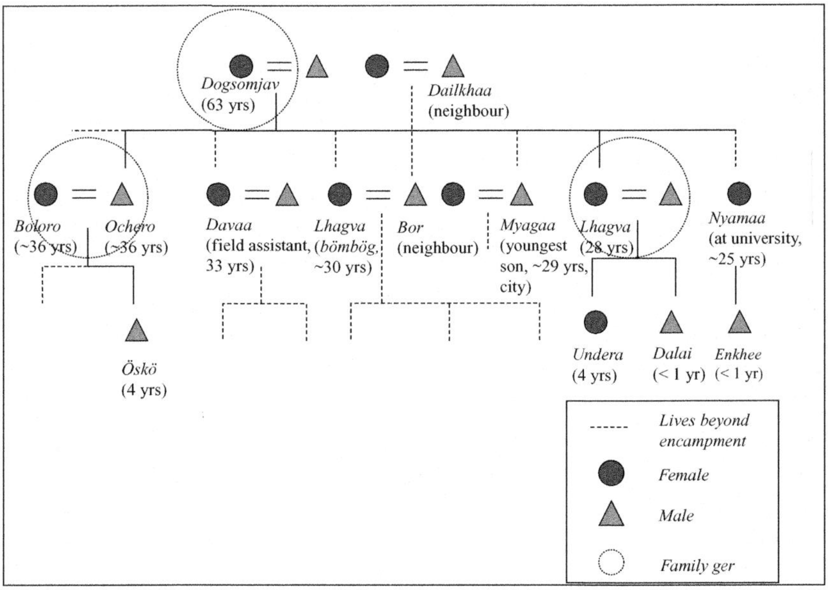

Figure P.1. Key people related to Dogsomjav's encampment, Bulgan, 2005.

offered invaluable assistance. The Agricultural University lay at the base of some low mountains, just beyond the outskirts of the smog-enshrouded capital. Under clear, blue, winter skies, I would escape the city to the comforting surrounds of academia. Inside the Ecology Department, covering an entire office wall was a map of Mongolia. I would discuss the logistics of potential field locations, using the huge map as a reference point. Tuvshin and Aagii proposed that I visit Aagii's father, who has been the local veterinarian in the area surrounding Bugat administrative district (*süm*) for the length of his working life. Aagii, like his father, trained as a veterinarian and was lecturing in the Department of Microbiology and Infectious Diseases at the University. In a break between lecturing duties, Aagii, Tuvshin, and I made a trip to Bulgan Province, in March 2005, so that I could see whether Aagii's hometown would be a good base location for meeting herders who would be open to my field research (see Map P.2).

We arrived in Bugat Süm and were welcomed in to Aagii's parents' home. This house was typical of many I visited in townships in the Khangai, consisting of a wooden exterior with thick concrete wall interiors, divided into two rooms, one with the fire and kitchen area and another for watching television and socialising but also functioning as the sleeping area at night. As with herding encampments, Aagii's extended family lived in two buildings next to each other. His eighty-eight-year-old grandmother and older sister with her children occupied one of these houses. His grandmother, puffing away on cigarettes, was remarkably lucid and fascinated as to why I would leave my home and my family to come to Mongolia. Her questions made me reflect on why I had actually done so. It struck me as a bit of a crazy enterprise when I viewed it from her perspective, as she had remained within the same village since her marriage seventy years earlier.

A local herder arrived unexpectedly, calling on Aagii's father to help him with a cow that had been having difficulty giving birth for the past two days. I took the opportunity to go along and witness the skills of a Mongolian veterinarian. We (the women of the family and I) watched helplessly while Aagii's father heaved on the calf's protruding legs. It was a painstakingly long time before he pulled the still-born calf out of the Mongolian cow, which looked very small in stature compared with the larger European breeds of cow I was used to.

It was when bumping along on the return to the village in the jeep, feeling disoriented by the darkness outside and the rough, snow-covered terrain, that I first met Dogsomjav, who happened to be visiting this same herding family. I could perceive that the herder driving the jeep was treating this husky-voiced woman in the passenger seat with great respect, discussing the cow's unsuccessful birth. Aagii announced to me that the older woman was Dogsomjav and that I should live with her encampment (*khot ail*) in the following months. Unlike many Mongolians who seemed unsmiling and guarded upon meeting me initially, Dogsomjav beamed, revealing a distinctive gap between her front teeth. She was wearing the typical attire of older Mongolian women: a brightly coloured scarf and a beautifully crafted, handmade coat (*deel*) with geometric-looking symbols on burgundy-coloured silk with a thick lining of lamb's hide for the cold of winter.

The next morning Aagii organised a jeep and a driver as a means of visiting some of the herding encampments. In early March the herders are still located at their winter sites. They are situated two bumpy hours of jeep drive up the broad river valley beyond the summer sites. The winter encampments are the longest distance from the local village (Bugat Süm) in comparison with other seasonal encampments. We followed the frozen Khujerin River Valley as the hills steadily became steeper and the valley more enclosed. The landscape was white apart from the exposed, windswept parts of the hills and the four-wheel-drive track we were following. On the mountain slopes bare-branched larch (in Latin *Larix siberica*, or Mongolian, *tsagaan khus mod*) and black pine (*Pinus sylvestris; khar mod*) grew in hardy clumps. The day was unusually warm with a north-west wind and a weak sun, which was enough to make the ice along the edge of the river slushy in places, making the going in the Russian jeep slow and treacherous.

Table P.1. Animal numbers at the Bulgan field site (per encampment)

Neighbouring encampments in Khujerin Valley (summer 2005)	No. of sheep/goats	No. of lambs/kids	No. of cattle	No. of cows with calves	No. of horses	No. of mares with young
Dogsomjav	76/73 (149)	47/45 (92)	32	18	34	7
Myagmaa	110/120 (230)	64/51 (115)	–	5	25	6
Dalkhaa	300/50 (350)	160/30 (190)	113	15	47	6
Sukhbaatar	(300)	–	–	13	36	10
Davaa	(400)	(200)	40	23	60	10
Demuul	200/100 (300)	(90)	60	22	38	7
Battogtokh	360/60 (420)	190/25 (215)	60	25	24	5
Myagmarsuren	22/18 (40)	0	15	5	4	0

Note: These numbers are herder estimates, not actual animal counts, and would be constantly subject to change. For columns representing more than one type of animal, total numbers are given in parentheses.

The first herding encampment we visited was Dalkhaa's. Four *ger* were situated on a sloping hill with a scattering of animal shelters for protecting the large sheep herd against the cold of winter (refer to Table P.1 for animal numbers). Dalkhaa loomed large within his *ger*, sitting with a shaved head and a rotund belly, looking like a wise lama. Dalkhaa had lived as a herder throughout his life and was a prominent person within the surrounding herding community during the Soviet era. He said to Tuvshin that he was pleased that a foreigner was interested in his herding knowledge and complained that veterinarians, like Aagii's father, didn't ask herders for their opinion enough. The members of this large extended family were later important as neighbours; I would often visit Dalkhaa and his thin, stately wife Dolgorsuren, or Bömbög (meaning "ball"), who married into Dalkhaa's family and is one of Dogsomjav's daughters. Dalkhaa was always willing to chat. He could no longer herd as a result of ill health but would while away the time by playing cards with neighbouring herders.

The next herding encampment was on the opposite side of the broad river valley that we were following. This was Dogsomjav's encampment, where I would stay in the following months. Dogsomjav was still visiting friends in the village centre but we were welcomed into the *ger* by Lhagva, her fourth daughter. Dogsomjav has ten children and twenty-six grandchildren. Four of these grandchildren were born during the year I was in Mongolia. Two of the children live in the encampment with Dogsomjav, namely Lhagva and Ochero, with families of their own (see Figure P.1).

Aagii's father joked that Lhagva's features were more Russian than Mongolian as she cooked us a meal of mutton on the bone and wide noodles, made laboriously by kneading flour dough and then dropping slices of the dough into the boiling broth. Lhagva did not look particularly Russian to me but she and her children had fair complexions. I could tell immediately that Lhagva was an engaging character, firing quick-witted jokes and talking animatedly.[5] Her brother Ochero was very different in character. He came striding into the *ger*, taking a midday break from herding the sheep and goats further up the valley. He gracefully kneeled on the floor, propping one of his long-sleeved arms on his raised knee, as is the posture of many herding men. He did not say a word. Lhagva, preparing the meal at the stove, held the conversation. Lhagva's baby, Dalai, her daughter Undera, and Ochero's son Öskö were the only children within the encampment during this time, as school-age children are generally sent to urban areas for schooling.[6] Although neither Undera nor Öskö are particularly shy, both were quiet in meeting me; like the rest of the family, they had not encountered a Western foreigner previously. I noticed that this encampment was very ordered, as the cooking implements and bowls were shining and spotless, and the area around the *ger* and animal pens was clear of animal dung. For the Khangai region, Dogsomjav's encampment herds an average-sized herd and has a good number of Mongolian cows for milking (refer to Table P.1 for animal numbers).

A neighbour happened to be visiting when we arrived and was sitting, sewing traditional Mongolian clothing (*deel*). Tuya lived near Dogsomjav's winter camp (approximately 500 metres behind the encampment) with her husband, Myagmaa, and their two adult sons in a small five-sided home (*ger*). During the colder months, Myagmaa and Dogsomjav's encampments combine their sheep and goat herds but keep their cattle and horse herds grazing separately. One of the adults from either family herds the sheep and goats every third day. Many herding families combine tasks with neighbours if the encampments do not have many inhabitants capable of herding. The partnership of two encampments for herding the sheep and goats is called *saakhalta*. In March, the female animals are not yet milked, so the

[5] See Lhagva the Herder, available at www.cambridge.org/9781107000902.

[6] Through the influence of Tibetan Buddhism, many Mongolian names are Tibetan in origin. Most of Dogsomjav's children were given Tibetan names for the days of the week. Two of the children were named Wednesday (*Lhagva*) but the elder of the two is always referred to by her childhood nickname of "Ball" (*Bömbög*). She is married to Dalkhaa's eldest son, Bor. Lhagva's baby son is named Dalai, as in the Dalai Lama, which means "ocean" in Mongolian (see Figure P.1).

cattle and horse herds were grazing on the sparse, dried grass stalks beneath a thin layer of windswept snow several kilometres from the encampment.

The families I visited were very welcoming. No doubt it had been helpful to be introduced by Aagii's father, who regularly visits the herders on his veterinary rounds. In winter he would visit on horseback but when the snow melted he would make his rounds by Russian motorbike. Tuvshin informed the families that we visited that I was merely considering the valley as an area for research; to my surprise, they said that they hoped I chose them. It was after this initial visit that I decided that this would be one of my two field locations (refer to Map P.2 for the Bulgan field site).

ARKHANGAI FIELD SITE: CHOIJO'S ENCAMPMENT

I went to Mongolia with the intention of establishing two field sites so that I could make a comparison between herd encampments and the differing structure of the herds. I planned to establish my two field sites in the area of the Khangai Mountains, containing forested slopes and grassland in the valleys. Through lecturers in the Anthropology Department at the National University of Mongolia, I arranged to visit a herding family who live in Arkhangai Province in March 2005. After an uncomfortable twelve-hour journey in a dilapidated bus with frost-encrusted windows, I met a relative of the family, who resides in a traditional *ger* but within Tsetserleg, the main township of Arkhangai Province. Homes in the urban *ger* districts tend to have the same layout as homes in the countryside but have the addition of electric light and perhaps a television. The following day we packed into a Russian jeep (including relatives and children anxious to visit for the day) and headed for Choijo's winter encampment. We followed an expansive river valley, passing herds of yak and yak–cow hybrids, in contrast to the Mongolian cattle inhabiting lower elevations.

Choijo's winter encampment is significantly higher in elevation (1,940 metres) than Dogsomjav's winter encampment (900 metres), although both environments are similar.[7] The families migrate up and down two broad river valleys within the Khangai Mountains. Both areas are relatively prosperous, with enough pasture for haymaking during autumn; the quality of pasture is good and herders are able to milk mares to produce *airag*. Both areas have patches of forest on the surrounding hillsides, which can be utilised for fuel

[7] At the lower Bulgan field site (Dogsomjav's location), the winter camp is at 900 metres; spring and autumn camp, 913 metres; summer camp, 1,182 metres. At the higher Arkhangai field site (Choijo's location), the winter camp is at 1,940 metres; spring camp, 1,905 metres; autumn camp, 1,910 metres; summer camp, 1,850 metres. Altitudes above sea level are obtained from global positioning system (GPS) positions on location.

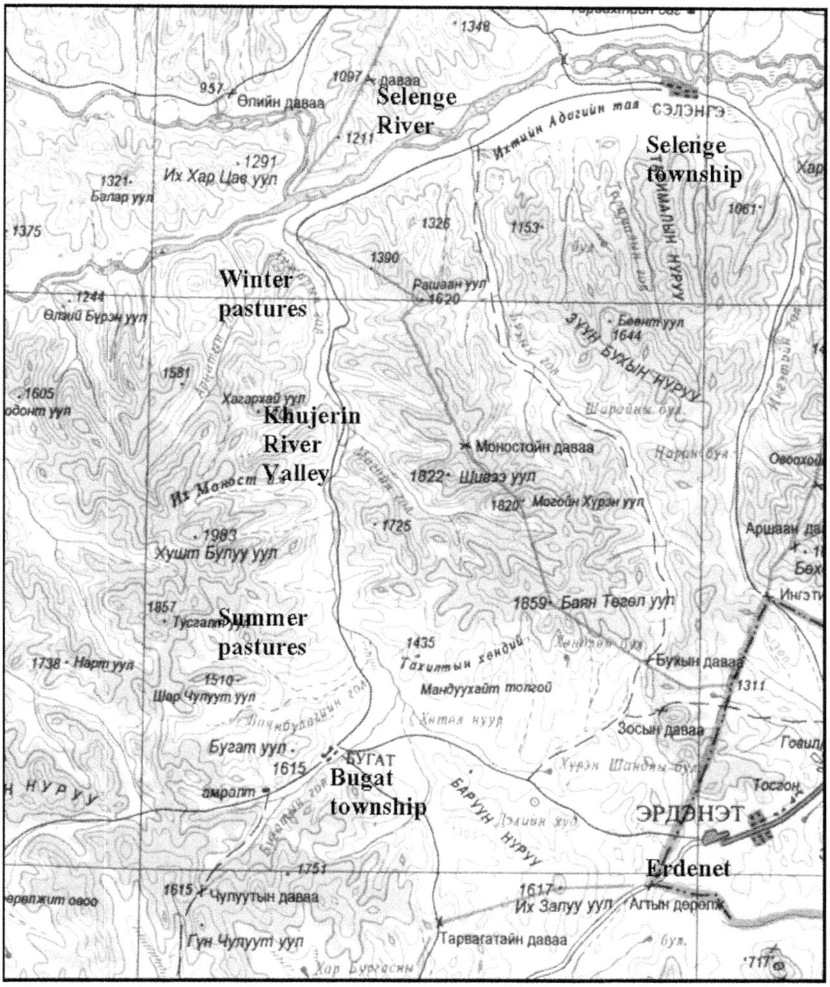

Map P.2. The Bulgan field site (scan from a Mongolian topographical map, *Zurag zui*, 2003; the scale is 1:500,000).

and building enclosures.[8] Because of its higher elevation, Choijo's encampment herds yak and yak–cow hybrids, whereas Dogsomjav's encampment herds Mongolian cattle.

Choijo's winter encampment was nestled at the foot of converging mountains, near stands of black pine forest, with a frozen stream near a line of four

8 In many grassland steppe and desert regions in Mongolia, the amount of standing fodder is not enough to make hay for supplementary feed; the quality of the grass is not nutritious enough to produce good-quality mare's milk; and the landscape is devoid of trees and therefore wood (Divangar, personal communication, Ulaanbaatar, November 2005).

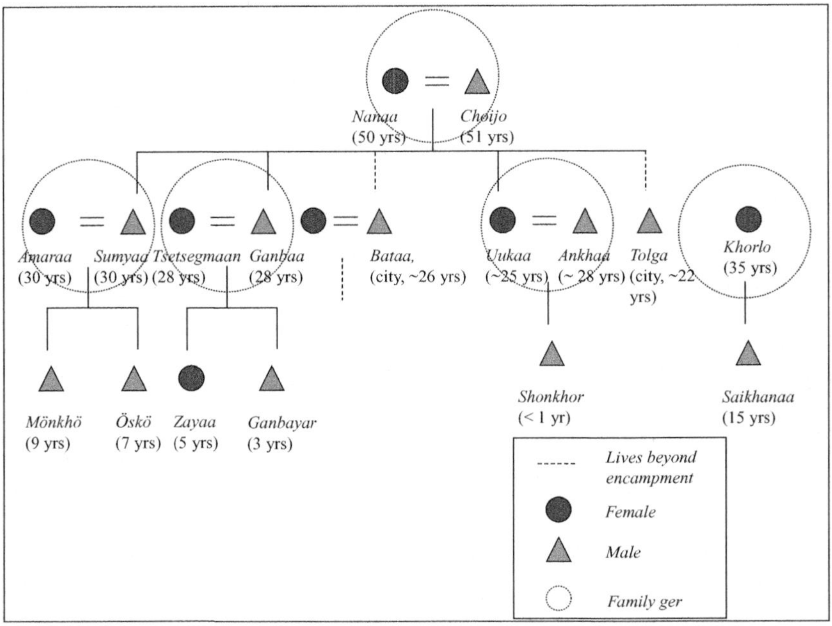

Figure P.2. Key people related to Choijo's encampment, Arkhangai, 2005.

ger. In this sheltered location, it felt noticeably warmer than did Ulaanbaatar, with its biting winds and icy pavements. Even so, such mountain regions in the Khangai ranges have air temperatures lower than −25° C for 100–105 days of the year with snow covering the ground for 120–150 days of the year (Bynie 2002). Upon this initial visit I stayed in Choijo and Naraa's *ger*, as they were my hosts, but later I lived in one of their smaller *ger* with my field assistant, Baigalmaa (see Figure P.2).

Naraa is a rotund motherly figure. She immediately made me feel at home with her warm welcome and rapid chatter as she worked preparing a meal. Choijo sat in the position of the senior male to the rear of the *ger*, in front of the small Buddhist altar and family photos. He occasionally acknowledged agreement but was content to puff away on his long-stemmed pipe, while Naraa spoke, hardly pausing for breath. Choijo's large seven-sided *ger* was positioned nearest the sheep and goat pen. A smaller *ger*, situated behind theirs, housed Khorlo and her son Saikhanaa.[9]

I was immediately interested in Khorlo's relationship to others in the encampment because she would enter Choijo's *ger* and sometimes smile as she listened to the conversation while bringing in firewood and stoking the fire, but she did not sit down, as everyone else did. I asked Naraa about the

[9] See Saikhanaa the Herder, available at www.cambridge.org/9781107000902.

relationship between her and Khorlo, and she said "she is my sister." At first I took this to mean that she was literally her blood sister, although Khorlo was physically very different, as she is small in stature and hunched over, with ruddy cheeks and worn hands from constant hard physical work in extreme cold. She looked far older than her thirty-five years. Mongolians, particularly in the countryside, address one another as relatives, for example "older sister," "brother," or "uncle," which is a formal but friendly means of addressing people who are not necessarily genetically related. It was only later that I realised that "Sister Khorlo" was essentially Choijo and Naraa's servant because she had no means of living by herself, with her only son Saikhanaa. When I asked a number of Mongolians about this living arrangement, they replied that it was not an unusual situation, as extended families that have large herds often require extra help. According to Bold, up until the 1950s, poor herding families were often connected with wealthy families: "When poor families were not able to subsist by themselves, due to insufficient quantity of stock, they were linked with the wealthy families both economically and privately" (Bold 2001: 71). Tuul commented to me that taking in a struggling family seemed to be more common since the economic collapse in Mongolia and could be a partial reversion to the so-called feudal structure that existed prior to the Soviet era.[10] If a herding family does not have a large number of children, relatives may send their own children to live with a household that needs assistance, and this herding family essentially adopts the child as part of the encampment.

Saikhanaa rode in to the encampment at dusk with the sheep and goat herd. At first I thought he was a boy of about ten or twelve years of age but later learnt that he was fifteen. His face was burnt a shiny red by being out in the extreme cold and he was shy in acknowledging my presence. It interested me that someone so young and slight in build could spend all day by himself, without food or extra clothing, herding the sheep and goats throughout the winter in temperatures that are often below −25° C. Some of the oxen had roamed further than usual that evening, so Saikhanaa went out on horseback well after dark to bring the oxen back to the winter site.

At this time of year, ewes and nannies were giving birth throughout the day and night, and during the initial days after birth everyone was assisting the newborns to feed from their mothers. Saikhanaa was rushing from mother to offspring and excelled in identifying which newborn belonged to a particular mother. He was careful when handling the young animals,

10 There is some debate, particularly in the Soviet literature (e.g., Balzer 1990; Vainshtein 1980) as to whether Mongolian society was in fact feudal. Historically, a commoner family associated with nobility was called *khamjaanai ail* (Vreeland 1953).

Table P.2. Animal numbers at the Arkhangai field site (per encampment)

Neighbouring encampments in Jargalant Valley (summer 2005)	**No. of sheep/goats**	**No. of lambs/kids**	**No. of cattle**	**No. of cows with calves**	**No. of horses**	**No. of mares with young**
Choijo	60/120 (180)	37/58 (95)	130	45	126	30
Sukhbaatar	20/30 (50)	7/16 (23)	45	23	54	8
Zundui	50/30 (80)	21/9 (30)	35	24	60	9
Ganbaatar	130/50 (180)	40/11 (51)	65	37	62	21
Namkhia	0/50 (50)	0/10 (10)	41	10	22	3
Bandikhuu	9/57 (66)	4/21 (25)	75	39	60	9
Mikhaa	67/115 (182)	40/32 (72)	66	30	27	9

Note: These numbers are herder estimates, not actual animal counts, and would be constantly subject to change. For columns representing more than one type of animal, total numbers are given in parentheses.

whereas to my eyes the other children seemed to be quite rough with them.

At the winter site there were two other *ger*, one occupied by Choijo and Naraa's only daughter, Uukaa, and her husband Ankhaa. Uukaa was seriously ill with kidney problems, so she was unable to participate in any work with the animals. She had been ill since the birth of her child and had been in the hospital most of the winter. Her husband Ankhaa got on well with Choijo and Naraa's third son, Ganbar. Naraa showed me photos of Ganbar and Ankhaa posing proudly in their neatly pressed army uniforms in the main square of Ulaanbaatar, in front of a statue of the hero Sukhbaatar, during their compulsory service in the Mongolian army. Ganbar lived with his wife Tsetsegmaa and their two young children within the encampment in the fourth *ger* (see Figure P.2).

Ganbar's family had a separate corral near their *ger* for milking the few cows still producing milk, but all the sheep and goats were combined as one herd. The horses were herded into the encampment once a day to check for injury, illness, or whether any mares were about to give birth. Otherwise the horse herd grazed out of sight of the herding encampment further up the valley. This encampment had a particularly large number of horses, probably because Choijo's father and later Choijo were allocated up to 400 horses at a time by the local cooperative (*negdel*) during the Soviet era. Choijo's encampment had more herd animals than any others in the Jargalant Valley (refer to Table P.2).

A few days after my arrival Choijo and Naraa's eldest son, Sumya, and daughter-in-law, Amaraa, arrived with their two children and Uukaa's baby, nicknamed Shonkhor. Sumya and Amaraa had temporarily moved to the nearest village during the winter so that their children could attend school,

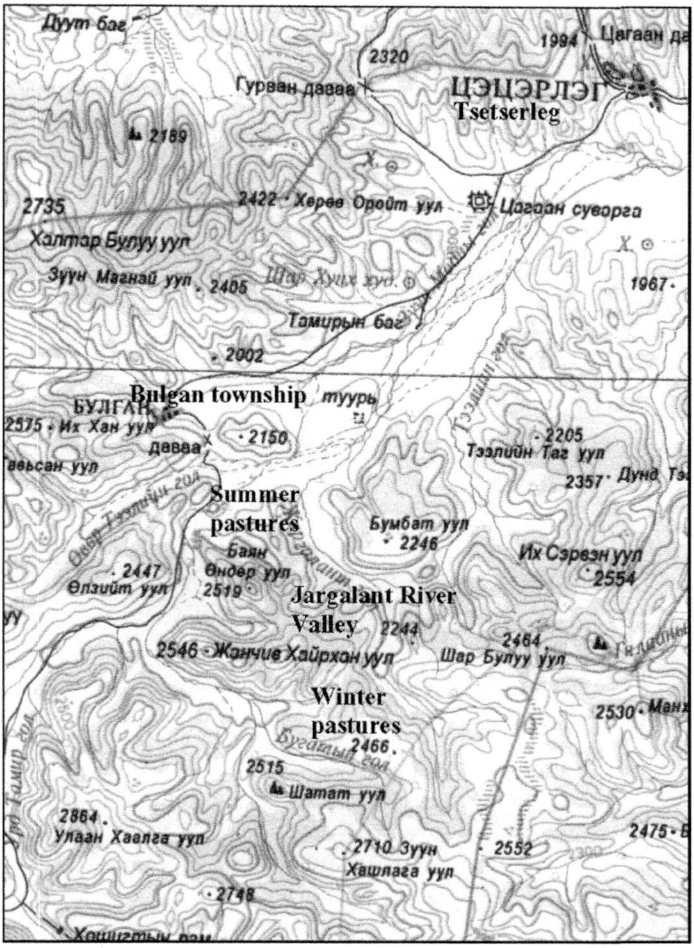

Map P.3. The Arkhangai field site (scan from a Mongolian topographical map, *Zurag zui*, 2003; the scale is 1:500,000).

and they had been caring for the baby during Uukaa's illness. Ankhaa, Amaraa, and Tsetsegmaa had all moved in to Choijo's encampment through marriage, but their parents and relatives live in neighbouring encampments within the same valley. Many of the people in the encampments within the Jargalant Valley are related through descent or marriage: Both Choijo and Naraa's parents had herded in the same valley; at the summer encampment, Choijo's sister and her extended family were close neighbours; and Naraa's brother and Amaraa's parents' encampments were within 100 metres of Choijo's autumn encampment (refer to Map P.3 for the Arkhangai field site).

During 2005 I was initially in Ulaanbaatar during the winter; I then lived at either Choijo or Dogsomjav's encampments. After the initial visits to both encampments, I returned again to Dogsomjav's encampment and lived there until the beginning of summer; the rest of summer until the start of autumn was spent at Choijo's encampment; and then during the rest of autumn until the start of winter I was again at Dogsomjav's encampment. I spent a roughly equal amount of time living at both extended family encampments and would come and go at times that I thought were most appropriate for each family, coinciding with seasonal moves to new pastures, or changes within the extended family itself. I visited both extended family encampments again on a short follow-up trip during the spring of 2007.

1

INTRODUCTION

The focus of this book is on the processes of animal domestication, not in a historical sense but within a current hybrid community, that of Mongolian herders and the herd animals they live amongst. In the chapters to follow I investigate social behaviour between humans and other animals as a key component in this process. I emphasise both sides of the human–animal relationship by examining their reciprocal social behaviour and communication with one another. I engage not only with how herd animals have an influence upon Mongolian herders' lives, but with the herd animals themselves as individual agents, as an active part of the community.

Herding is a way of life for a large proportion of the people in Mongolia and has existed for over 5,000 years in the Eurasian grassland–steppe environment. Anthropologists, geographers, historians, explorers, journalists, and travel writers have previously documented Mongolian society; nonetheless, I am unaware of any other work written in English that has examined the importance of herd animals, specifically ungulates, to nomadic pastoralists of Mongolia. In this chapter, I describe what the term "domestication" means to Mongolian herders and within the academic sphere, as well as my own definition of a domestic animal for the purposes of this book. I provide a background to the literature on the domestication of the animals that Mongolians herd and a brief historical account of herding life in Mongolia. I also characterise the literature that has been influential in forming my ideas in relation to my etho-ethnographic research, particularly the human–animal related, cross-disciplinary work of academics such as Tim Ingold (1986a, 2000), Donna Haraway (2003, 2004, 2007), and Jared Diamond (1997). Ethnographies on other peoples of Inner Asia and Northern Asia have also been influential to my approach, such as the work of Caroline Humphrey (1983; Waddington-Humphrey, 1974), Piers Vitebsky (2005), and again Tim Ingold (1980), and their work is discussed in the paragraphs that follow.

PART I. REDEFINING DOMESTICATION

The domestication of an animal by humans, forming and making the animals into a being of human design, is likely to be a Western concept. There are words in Mongolian that can be translated as applicable to so-called domestic animals, derived from the animals being part of a domestic sphere (*geriin tejeever*), or associated with the home (*ger*). Mongolians refer to the animals that they herd simply as herd animals (*mal*), or the five kinds of animal (*tavan khoshuu mal*), and to themselves as herders (*malchin*). Wild animals, such as deer or bears, are categorised as animals (*am'tan*) rather than herd animals (*mal*). These animals have no contact with and are afraid of humans, so they are therefore "wild" (*zerleg*). "Domestication as a process raises intriguing issues of intention, selection and co-evolution, and poses in an acute form the difficulty of the terms nature and culture in scientific discourse" (Ellen & Fukui 1996: 21). Just as there has been a movement to redefine nature in relation to culture in anthropological discourse, my premise is to argue for a redefinition of animal domestication as a term and what it means for an animal to be "domestic" or "wild."

Sandor Bökönyi defines animal domestication in the following way: "The essence of domestication is the capture and taming by man of animals of a species with particular behavioural characteristics, their removal from their natural living area and breeding community, and their maintenance under controlled breeding conditions" (Bökönyi 1989: 22). O'Connor (1996) convincingly argues that this definition of the domestication process is heavily one sided, with the animals passive in the process, moved around and controlled for human ends. He states, "much published work on animal domestication is valid but much is constrained by a strongly anthropocentric belief that the human:domesticate relationship is something unique, to be explained by reference to unique attributes of human socio-economic behaviour" (O'Connor 1996: 155). He proposes instead that domestication should be considered in biological terms, as a symbiotic relationship.

Jarman et al. (1976) prefer to use the term "animal husbandry" but still emphasise the regulation, manipulation, and exploitation of the animals, rather than a more reciprocal relationship between human and other animals. Pierre Ducos also defines domestication from an anthropocentric viewpoint when he states that "living animals are integrated as objects into the socio-economic organisation of the human group" (Ducos 1989: 28). This definition refers to the domesticated animal as an object or thing rather than as a person or subject. As I will argue later in this book, Mongolian herders do not view the herd animals as objects, or wholly as economic commodities, but as "persons" with individual characteristics and personalities.

In other words, Mongolian herders have a different perspective and attitude towards their herd animals compared with most views espoused in Western discourse, where an animal is perceived as an economic resource that can be made by humans for their own consumption.

Tim Ingold's definition is relevant to my perception of domestication, with his emphasis on social rather than physiological processes. Ingold writes in relation to reindeer, "I do not mean selective breeding towards a form that is physiologically dependent on man, but the element of socialisation of the animal into a human environment" (Ingold 1974: 524). He emphasises this socialisation as an integration into human society, although it is my view that non-human animals are active agents in this process, as humans reciprocally become integrated into herd society (see particularly Chapters 6 and 8). Lestel (1998) refers to domestication as not being about the state of the animal but the interaction between human and animal within an integrated set of socio-cultural structures. Lestel emphasises that domestication is both mutual and cognitive, developing specific expertise in their relationship with one another. Haraway (2003) avoids the use of the term "domestic animal" and instead refers to "companion species," but this type of companion relationship exists more in a one-to-one partnership with a particular dog, horse, or cat and is not as applicable to whole herds of animals, as is the case in Mongolia.

In the chapters to follow, I refer to the Mongolian relationship between herders and herd animals as being "co-domestic," in contrast to the intensive, consumer-driven animal husbandry techniques used with "domestic" animals on Western farms (see Bulliet 2005; Noske 1989). Anderson (1986) uses the term "codomestication" to describe the relationship between Saami reindeer herders, dogs, and reindeer but relates this mainly to phylogenetic adaptation rather than behavioural adaptation processes. My definition of a co-domestic relationship is the social adaptation of animals in association with human beings by means of *mutual* cross-species interaction and social engagement. I intend to avoid the past connotations of domestication, in which the animals are necessarily restricted within a captive environment, or the reference to animals as economic commodities, purely under human control. I demonstrate in subsequent chapters how Mongolian herding aligns with this definition of a co-domestic relationship, rather than the anthropocentric definitions of domestication that have dominated the animal domestication literature. In Mongolia, herders and herd animals live with each other in a shared landscape, inhabiting a co-domestic, ecosocial sphere: the herding encampment (*khot ail*). In my view, which I believe aligns with Mongolian herders' views, it is the degree of social interaction and engagement with humans within this ecosocial sphere that dictates whether an animal is a co-domesticate or not.

ORIGIN THEORIES

Previous literature on domestication has tended to argue that the domestication of an animal was through intentional invention by a small group of people, or even one individual, followed by subsequent cultural transmission of the idea (see Serpell 1989). This view of animals as passive in the initial establishment of domestication promotes the view of domestication as a feat of engineering, as though the ox were man-made, an artificial construction put together like the plough (Ingold 1994, 2000). In other words, this follows the idea that the domestication of nature gave rise to culture, or the "cultivation" of nature. A relative of Charles Darwin, Francis Galton, was an advocate of the theory that animals were originally domesticated from orphaned animals that became pets. David Rindos proposed a different theory, that domestication was not invented by humans but that domestication occurred as part of a process of natural selection, a co-evolution, even symbiosis between humans and either plants or animals (Rindos 1980; see also Zeuner 1963).

Communication between different species by no means arises exclusively from the human's intentions and intelligent design. A good example of a symbiotic relationship between humans and animals is the Honeyguide bird (*Indicator indicator*), who shows individuals from the nomadic Boran tribe in Africa the location of a beehive through communication of specific vocalisations and by maintaining a particular flight pattern just ahead of the humans, until they reach the beehive, whereas in return the Boran summon the bird through special whistles (see Hooper 1989). The human participant benefits in this relationship from a source of honey, and the Honeyguide benefits from feeding on larvae once the nest has been cut open. The bird has not been tamed or trained by humans but both parties have mutually adopted the cooperative strategy over time. It is important to note that zoologists have focused more on biological interspecific associations (such as mutualism and symbiosis) in the past but less on behavioural interspecific associations (see Lestel, Brunois, & Gaunet 2006).

Scientific evidence of domestication in a novel animal species has shown that a key factor in the domestication process is the selection of a specific *behavioural* trait: tameness. Over many generations, animals closely associated with humans have been (consciously or unconsciously) selected for tameness, calmness, and placidity. At the Institute of Cytology and Genetics in Russian Siberia, scientists have conducted a longitudinal domestication experiment over a period of forty years on the Silver fox (*Vulpes vulpes*, which is in the same family as the domestic dog). The director of the institute, Dmitry Belyaev, developed what is now called Belyaev's Hypothesis, which is that domesticated animals are often selected primarily for behavioural traits and that subsequent physiological and morphological changes are a secondary result of this selection for behaviour (Belyaev 1979; Belyaev, Plyusnina, & Trut 1984/1985). Through intensive selection, the domestication of the Silver fox has undergone a process within a few decades that would have spanned thousands of years under normal environmental conditions. After thirty to thirty-five generations of selecting foxes specifically for tameability, strong phenotypic domestication traits were revealed, such as curly tails, piebald and star markings, and floppy ears (Trut 1999). The Institute also found that the eventual tameness of the foxes in the presence of humans was directly related to a change in brain chemistry in the hypothalamus (Price 1999). A complex system of genes (polygenes) not only controls the functioning of the neural and endocrine systems but can also govern an animal's behaviour, including tameness or defensiveness towards humans. The foxes reacted to humans in a similar manner to dogs in their attachment and sociability towards them. This experiment clearly shows the importance of the selection for behavioural traits in the domestication process, particularly tameness towards and bonding with humans (for more on the importance of "tameness" in Mongolia, see Chapter 6).

Cognition-based experiments that were originally developed for preverbal infants have provided evidence of the adaptation of dogs to methods of human communication through the process of domestication. The experimenters pointed or merely gazed at a food bowl and tested the responses of both dogs and wolves. Wolves were not able to follow human gaze to a food source, whereas dogs could do so as puppies, even though both the wolves and dogs were habituated to human contact. Highly social animals are likely to have evolved mental processing abilities to help them predict and exploit the actions of others. Somewhere during the process of dog domestication, dogs that could follow the gaze of humans fared better than those that could not interpret this method of social communication (Cooper et al. 2003; Hare et al. 2002).[1] In additional experiments but this time testing the cognitive skills of goats, ethologists found that the goats were also able to follow human gaze to locate food (Kaminski et al. 2005). Here it is evident that, through constant interspecies interaction and communication, the actual behavioural traits exhibited in domesticated animals can change through gradual adaptation for reading human cues and signals. Within this

[1] Much of the literature on human–animal relationships and domestication processes have focused on dogs (see, e.g., Serpell 1995), particularly as companion animals, but has a tendency to neglect other animals. The Mongolian dog's role is as a guard dog within an encampment but such dogs are not kept as pets or for herding animals (for more details on the role of the Mongolian dog, see Chapter 8).

book I aim to investigate how the interactions between humans with other species, such as horse, cattle (including yak), sheep, and goat, influence herd sociality, not only of the species of the ungulates in question but also their human counterparts (in particular, see Chapters 5 and 6).

Donna Haraway, in *The Companion Species Manifesto* (2003), discusses the relationship between companion species, specifically dogs and humans, as a co-evolutionary process. This aligns with the idea that it is not just humans who chose to domesticate animals but that the relationship worked because the animals were also agents in the process. She writes that companion species mutually adapt. An undeniable aspect is the sharing of pathogens, which in turn alters immunity, "but it would be a mistake to see the alterations of dogs' bodies and minds as biological and the changes in human bodies and lives, for example in the emergence of herding and agricultural societies, as cultural, and so not about co-evolution" (Haraway 2003: 31). She reemphasises this co-adaptation in her more recent work stating that we all originated symbiogenetically from the earliest nucleated cells: "I believe that the *embryonic* co-construction of the physical bodies has many more implications because it means that we were 'never' individuals" (Haraway 2007: 34). Haraway has altered the phrase coined by Bruno Latour (1993), "we have never been modern," to indicate that we have never been human either. Anna Tsing agrees that species' interdependence is a well-known fact, except when it comes to humans: "Human exceptionalism blinds us" (Tsing in press). She too refers to human nature itself as an "interspecies relationship." Like Haraway, I became interested in researching animal domestication because I wanted to investigate whether this domestication was a co-evolutionary, symbiotic process rather than the predominant Western perception that it involves a passive response from the animal, based on human intention and design.

Animal domestication is often investigated in the realm of archaeology, especially zooarchaeology, in an attempt to establish the origin of domestication in animals. This line of questioning could be flawed from the outset in that the aim is to seek a point where humanity transcended nature and was set along the path towards culture and civilisation. Anthropological thinking is moving towards the idea that humanity has never been separated from nature, and this same line of thought can be used in relation to non-human animals (Ingold 2000). Just as it is impossible to discover the exact point at which one species adapts, or evolves, into another, there is no one point at which a species changes from a state of being wild and part of nature to a state of domestication and part of culture. A species of animal is not wild for one generation and domesticated the next, as the process is an *ongoing* mutualistic relationship in which both partners become interdependent (Shanklin 1985; Zeder et al. 2006).

Identifying whether animals at human sites were domesticated or wild is often a problematical process, as it is reliant on the bones of the domesticated animals being somewhat smaller than wild representatives, a process of paedomorphosis; a predominance of juvenile bones consumed by humans at the site; or on a change in the animal's skeletal morphology or dentition (see, e.g., Anthony 2007; Levine et al. 2000). New techniques of DNA analysis have clarified a lot of the questions surrounding the origins of domestication. It is evident, through tracing both patrilineal (Y chromosome) and matrilineal (mitochondrial DNA) heritage, that animals such as dogs, horses, and cattle have originated from a number of separate locations (Bruford, Bradley, & Luikart 2003).

A fascinating aspect of the domestication process is that animals not only change in physical appearance but also change behaviourally. Behavioural traits are sometimes linked to physiological traits, but behaviours have the potential to change very rapidly, on a separate time scale from changes in taxonomy. From the time of Darwin's *The Variation of Animals and Plants under Domestication* (Darwin 1868), research on animal domestication has often neglected the selection of behavioural traits as an important factor in the domestication process and instead has tended to focus on the outward changes in physical characteristics. This emphasis is not surprising, as it is relatively easy to compare the physical differences between a Great Dane and a Miniature Poodle, but somewhat more demanding to characterise their behavioural differences. The only way to examine behavioural changes in domesticated animals is to examine present-day populations to gain a window into the past.

UNGULATES IN MONGOLIA

One reason why I chose to focus on ungulates (or mammals with hooves) is that they are essential to the nomadic pastoral existence. I began to engage in research on domestic animals with ideas similar to those of Tim Ingold. He states that domestic animals, such as reindeer, "should engage our attention not merely as curiosities but because they furnish a crucial testing ground for some of our most general theories concerning the role of animals in human social evolution" (Ingold 1986b: 5). Mongolians herd four of the most successful domesticates: horse, cattle, sheep, and goat. Jared Diamond (1997) regards these as part of "the major five" (as he also includes the pig but Mongolians include the camel). I chose the central region of Mongolia, the Khangai mountain steppe, as a broad field area because I noted that the density of ungulates is higher in this area, probably as a result of the better quality of pasture and more established infrastructure for trading purposes than that of other regions (Humphrey & Sneath 1996b: plates 1–4).

The large ungulates of the Eurasian steppe could not have easily co-existed with humans if the social behaviour of these animals had not been compatible with human social behaviour. Goats and sheep were particularly well adapted to coexistence with humans because they had a shorter flight distance than many other herbivorous mammals, which was due to their existence on steep, mountainous terrain. Their predecessors were able to retreat up steep slopes to escape from predators. Ungulates that coexist with humans have extended far beyond the range of habitat that they would have occupied in the "wild" and are in far greater abundance than their counterparts within the same genus.

In the prehistoric past, aurochs, yak, reindeer, camel, horse, mountain sheep, and goat lived on the Eurasian steppe and mountain taiga. These large herbivores had considerable and lasting impact on the landscape of the steppe throughout Eurasia. Khazanov (1994) comments that herds that were mono-specialised, such as reindeer and yak, were destined to remain primarily within their original habitat. Multispecies herds, such as those that occur in Mongolia, had a greater capacity to expand beyond their original locations across the Eurasian steppe. In contrast to most countries throughout the world, there are still representatives of most of these ungulate genera persisting in Mongolia.[2] Przewalski's horse (*Equus przewalskii*), or *takhi*, has been successfully reintroduced into Mongolia. Remnant populations of the Bactrian camel (*Camelus ferus*) still remain, as do ibex (*Capra sibirica*) and argali sheep (*Ovis ammon*).[3] To Mongolian herders, these animals are "spirit animals" containing great power, particularly the *takhi* (this is also true of the progenitor of the dog, the grey wolf, *Canis lupus*).

Unfortunately, all of the ungulates are critically endangered from overhunting in Mongolia; this is a worldwide trend for "wild" representatives in the presence of competing domestic animals. Farmers or herders and domestic stock gradually push wild members of the species into marginal habitats through occupation of prime habitat. The remaining wild animals often become genetically integrated with domestic stock through interbreeding, or killed off by hunting pressure if they encroach too close to the domestic sphere. These factors are likely explanations for the extinction of the progenitor of the cow, the aurochs (*Bos primigenius*; see Clutton-Brock 1987).

[2] For both yak and reindeer, Mongolia is now the outer limit of their natural range because there is less high-altitude taiga forest than there previously was. However, "wild" representatives of reindeer still exist in the Siberian Arctic and those of yak exist on the Tibetan Plateau (Heptner, Nasimovich, & Bannikov 1961; Schaller 1998; Zhang 2000).

[3] *Argali* is the Mongolian name for mountain sheep, *yangir* for mountain goat, *takhi* for the wild horse, and *khavtgai* for the camel. Because Mongolians refer to the wild horse as the *takhi* and not by the European term, Przwalski's horse, I will henceforth refer to this horse as *takhi* and use the term "Mongolian horse" for those that coexist with humans. Note that in Mongolian the "i" on *takhi* and *argali* is not articulated.

In the remainder of this section, I briefly outline the history of the horse, cattle (including yaks), sheep, and goats in relation to Mongolia, as these are the ungulates found in the mountain–steppe habitat of the Khangai.

Mongolian Horses

Mongolian horses are small in stature, "well-muscled, thickset, short-legged and heavy-headed, resembling the wild *Equus przewalskii*" (Osmaston et al. 1994: 223). They are able to survive through harsh winters by developing a long, shaggy coat, and they paw through snow to forage on the sparse vegetation below. Mongolian horses are particularly hardy and have good endurance; although they are not large horses they are capable of carrying loads equal to one-third of their body weight (Tumerjav 2003).

Early depictions of the horse, in cave paintings found in Europe, are of a similar type to the *takhi*. Predomesticated horses were small in stature, had manes that stood on end, leg stripes, and a dorsal stripe along the back (Clutton-Brock 1992). Charles Darwin (1868) perceptively recognised these features as possible "wild-type" characteristics from an examination of various horse breeds. Occasionally, I saw some individual Mongolian horses with wild-type markings: horses that were dun in colour, with a stripe down the back and faint stripes on the neck and legs reminiscent of the *takhi*'s coat colour and markings (see also Hilzheimer 1935; Lusis 1941). The herders who rode these horses told me that they were highly prized for their colouration and distinctive markings.

Many Mongolians believe that the modern-day horse originated in the region of Mongolia and that the Mongolian horse was derived from the same ancestor as the *takhi*. Herders refer to the *takhi* as the "father" of the horse ("*mor*'"). A respected Mongolian academic states, "Mongolian horses are one of the oldest breeds of horses in the world. The closest ancestor of the Mongolian domestic horse is the wild horse known as *takhi*" (Tumerjav 2003). The *takhi* and Mongolian horses can interbreed and are capable of producing offspring that are fertile in both sexes (Bökönyi 1984). This could mean that Mongolian horses have had some *takhi* characteristics introduced through the male line. Mongolian herders have experimented with the hybridisation of yaks with Mongolian cattle for centuries, so it would not be much of a leap for herders to experiment with the hybridisation of equids.

There is DNA evidence to suggest that the horse was "domesticated" independently in a number of different locations across Eurasia (Pennisi 2001; Vila et al. 2001).[4] The horse has a high degree of genetic diversity with

[4] Mitochondrial DNA or mtDNA is ideal for studying the divergence of populations during a relatively short time scale (<10, 000 years). Analysis on the Y chromosome also provides important information on patrilineal inheritance.

phylogenetic clusters of mitochondrial DNA specific to northern European ponies, Przewalski's horses (or *takhi*), and northwest African breeds (Jansen et al. 2002). Significantly, the Mongolian breed of horse is closely associated with northern European horse breeds. Through tracing horse lineages with DNA and work on morphology by Colin Groves (1994), it is generally accepted that the wild *takhi* are not direct ancestors of the European breeds of modern-day domestic horses, because they have a different number of chromosomes (although, as already noted, there have been instances of hybridisation with Mongolian domestic horses). Bjørnstad, Nilsen, and Røed (2003) suggest that the Mongolian horse had a major impact on a wide range of breeds in northern Europe and that Central Asia, as a region, could have been the centre for horse dispersal through human trade and migration. The relationship between human and horse was undoubtedly established in Mongolia at a very early stage (also see Chapter 7 for archaeological finds of early horse sacrifice on the western border of Mongolia).

Mongolian Cattle

Mongolian herders have indigenous breeds of cattle and yaks that are particularly woolly and hardy in extreme winter conditions. Mongolian cattle are smaller with narrower hindquarters and produce less milk than most cattle breeds, but they are survivors, developing a thick coat and hide during the colder months of the year. This enables them to endure the extreme cold, even when the ends of their ears and horns have dropped off from frostbite. Present-day Mongolian cattle are well adapted to grassland–steppe habitat but are unable to survive at higher altitudes or in a desert environment.

Mongolian yaks (*sarlag*) are slightly larger than Mongolian cattle, with a curtain of hair on their underside and large, bushy tails, which they expressively wave about when excited, or agitated, accompanied by low, resonant grunts. The shape of a *sarlag*'s lips, tongue, and teeth allow it to forage on low-lying, scattered vegetation. They pant with a purple tongue hanging out when overheated. *Sarlag* are particularly hardy in high-altitude environments and may be ridden during the winter instead of horses, when horses become weakened by extreme weather conditions and lack of food. All of the stock that are native breeds to Mongolia have the ability to paw through snow to eat the fodder beneath, but sarlag are particularly adept at doing so. As Phillips and colleagues note regarding yaks in caravans in Tibet, "When stuck, the yak turns and tries another way, or may back out and start anew. When sinking, it walks in a long, slow stride, dragging its wide body along the swamp like a boat" (Phillips, Tolstoy, & Johnson 1946: 167).

In the higher mountain–steppe environment, Mongolian herders tend yak or yak–Mongolian cattle hybrids. Yaks are used extensively in the mountainous areas of Inner and Central Asia, such as the Pamirs in present-day Kyrgyzstan, through to the Altai Mountains bordering Mongolia and Kazakhstan, in the Khangai Mountains of Mongolia, and just to the North in the Russian Republic of Tuva (Ekvall, 1968; Jianlin et al. 2002). Mongolian yaks differ from the Tibetan yak breed and those in the Khangai Mountains differ slightly from those in the Altai Mountains. One such difference is that 90 percent of yaks in the Khangai Mountains are hornless. According to Heptner et al. (1961), there is archaeological and historic evidence that there were "'wild'" yak in the area of Tuva and the Altai mountain region on the borders of Mongolia, so wild yaks were therefore likely to exist in Mongolia too. Although the yak was more widespread in distribution in the past, wild yaks have become confined to only remote regions of the Tibetan Plateau through hunting pressure and loss of habitat.

Mongolians have a specific word for yak, *sarlag*, but they refer to both Mongolian cattle and yaks under the generic name for cattle, *ükher*. Herders favour yak–Mongolian cattle hybrids because they live longer and are capable of living at lower altitudes. The females produce more milk and young than pure-bred yaks. Male hybrids are placid and make good oxen, as they have a great capacity for pulling heavy carts.

Mongolian Sheep and Goats

Mongolia has a remarkable diversity of livestock breeds, most of which are indigenous to Mongolia, each breed having adapted to a specific region. This diversity is particularly evident in the sheep and goat breeds. Sheep from Mongolia are derived from different lineages from those found in Europe. Mongolian sheep have similar features to Tibetan long-wooled sheep, suggesting that both stemmed from the mouflon of South Asia, first arriving in Tibet and then dispersing throughout Central Asia (Tumerjav 2003).

To the casual observer, mountain sheep and goats look very similar, with large, curving horns and similar variations in coat colour and markings. (In contrast, the diverse breeds of modern Western domestic sheep and goat look markedly different from one another). I initially mistook newborn Mongolian lambs inside a *ger* during the winter for goat kids, and at first I could not believe that they were indeed lambs. The young sheep and goats have physical characteristics similar to one another and are more similar to "wild" representatives (for example, Mongolian sheep wool still sheds in summer). Mongolian sheep and goats grow more slowly than breeds in other regions of the world, ceasing any growth during the harsh winter and spring. Both lose as much as 30 percent of their body weight during the winter (Tumerjav 2003). Herders in Mongolia always keep the two species together, as a sheep and goat herd (*bog*), because they complement one another in their flock behaviour (see Chapter 3).

Herders view the herd animals as originating from "father" species of horse, camel, sheep, and goat that still roam the more inaccessible pockets of Mongolia today. The herd animals are seen as part of the herders' family, and the "wild" equivalents are seen as part of the extended family network. Kaczensky agrees when she states that "ibex (*Capra ibex*), and argali (*Ovis ammon*) resemble domestic sheep and goats and Przewalski's horses (*Equus przewalskii*) resemble domestic horses in appearance and behaviour. Like domestic stock these animals are viewed with great passion and seem to be regarded as the wild relatives of the extended human-livestock family" (Kaczensky 2007: 326). Mongolian herders view the animals they herd as an inherent part of the landscape. In Chapter 3 I show how herd animals exhibit similar social structures to those found in wild or feral populations. Instead of considerably altering herd behaviour, herders coordinate their own behaviour to complement that of the herd animals.

PART II. INNER ASIAN NOMADS

Mongolia is an independent country the size of Western Europe, wedged between two superpowers, Russia and China. More Mongolians actually live within the broader area of Inner Asia but outside the borders of Mongolia itself. Many live to the north in the bordering provinces of Tuva and Buryatia within Russia and to the south in the provinces of Inner Mongolia and Xinjiang within China. Mongolia itself is sparsely populated with only 2.6 million human inhabitants who herd an estimated 26 million herd animals (Johnson et al. 2006). The herding families and their herd animals occupy the largest area of common grazing land in the world.

With an average elevation of 1,560 metres above sea level, the air in Mongolia is rarified and the sky often an azure blue. A major part of the country is hilly, mountainous, or mountain–forest steppe, with a predominance of larch and pine forests covering the mountainsides. This book is based on field research from two locations in the Khangai mountain region of central Mongolia. Animal husbandry and herding practices are broadly similar throughout Mongolia but do inevitably differ depending upon habitat and environmental conditions. The high-altitude Khangai is ideal habitat for the yak, whereas it is unsuitable for the Bactrian camel, which thrives on desert–steppe vegetation; horses tend to be more abundant on the predominant grassland steppe.

Throughout this book, I use specific terms, such as "Mongolian," "herder," *ger*, and "encampment" (or *khot ail*; refer to the Glossary). Most of the herding people in the Khangai region belong to the ethnic majority, Khalkha Mongolian. Herders within the country of Mongolia refer to themselves generically as *Mongol*, or Mongolian. Mongolians who herd animals are encompassed under a variety of different terms in the literature relating to Mongolia, such as "pastoralists" (Barfield 1993); "mobile pastoralists" (Humphrey & Sneath 1999a, 1999b); "pastoral nomads" (Khazanov 1994); "nomadic pastoralists" (Goldstein & Beall 1994); or "herdsmen" (Lattimore 1962). These terms tend to emphasise the process of movement, the importance of pasture, or are specific to one gender. Humphrey and Sneath (1999), in their book *The End of Nomadism?*, question whether nomadism is any longer analytically useful as a term, as it has been so contested in the anthropological literature. My emphasis in this book is neither on nomadic movement nor on the use of pasture; it is on the social engagement with the herd animals themselves. I use the term "herder," which can mean any man, woman, or child who takes the role of directing the animals they herd. Mongolians refer to their way of life as *mal süreg*, or "to herd," and to themselves as *malchin*, "herders." I naturally adopted their terms while in the field, so continue to do so here. I use the terms "herding" and "taming" in a slightly different manner to Ingold (1980), as I refer to "tameness" in reference to a behavioural characteristic of an individual herd animal whereas "herding" is in reference to the leadership role of the herders, as a means of controlling the herd as a whole. I refer to the "herding family" (*ail*), which means a family group, typically encompassing a husband and wife and their children who usually occupy one tent, or *ger*, which is essentially home.

Caroline Humphrey's work on the use of horse brands in Mongolia is the ethnographic research that is particularly relevant to the research within this book (Waddington-Humphrey, 1974). Mongolian social organisation is similar to other herders on the Eurasian steppe. The basis is a patrilocal stem-family system, whereby the youngest son inherits the family herd from his parents after portions of the herd have been distributed to elder brothers (but now also sisters) when they marry. Humphrey (Waddington-Humphrey, 1974) recognised that a family's lineage could be interpreted through the branding marks on horses, or *tamga*, as the older brothers would adopt a symbol that was a slight variation from the family's symbol, while the youngest brother would adopt the same symbol as his father. The symbol also denoted whether a family came from an important lineage, such as a khanate, or whether the herd was linked to a Lamaist temple, as brands often depicted important Tibetan Buddhist symbols. These branding symbols functioned as visible markers, so that any herder riding across the grassland steppe was able to identify not only the relationship of a herd to a specific family but also the genealogy and lineage of the family, and even designate a family's role in society (see Chapter 4 for an account of the use of the *tamga* by herders in the Khangai Mountains today).

Prior to the Soviet era in Mongolia, herders developed animal husbandry and herding techniques through a process of learning from their parents and grandparents and by trial and error during a lifelong association with the herd animals, a process referred to in Mongolia as the "folk" (*arat*) method (see Chapter 4). This form of animal husbandry changed after 1921, as Mongolia adopted a form of socialism similar to that of the Union of Soviet Socialist Republics, or USSR (for an example of Soviet-style pastoralism in Buryatia, see Humphrey 1983). Mongolia relied heavily on Moscow for aid, and government policies were closely linked to those in the USSR. Herding households were formed into collectives (or *negdel*) to manage state-owned livestock. From the age of seven, all Mongolians were required to leave the countryside and attend school in the administrative centres that were established under the socialist regime. Every administrative centre, or *süm*, in the Mongolian countryside was allocated a local veterinarian. This inevitably changed the lives of the herders within Mongolia and altered their animal husbandry techniques. Russian ethnographers, probably as a result of Mongolia's close political relations with neighbouring Russia and the strong Soviet academic influence, have conducted most of the existing foreign research on Mongolia (see Khazanov 1994 for a review of the Russian literature on pastoral nomads in Mongolia).

From 1924, the Mongolian government began introducing exotic breeds into the state-owned *negdel*, particularly breeds from the USSR. They began introducing so-called modern animal husbandry practices (such as artificial insemination in 1945). Veterinary technicians, often trained in Russia, would advise herders on the number of animals to be produced and introduced breeds from Russia that were thought to increase productivity (Batsukh & Zagdsuren 1991). The herders could no longer make choices about their herds for themselves but had to conform to the government policy of the time, administered through local zootechnicians. It is evident from research into the number of animals from the 1920s through to the 1990s that government policy did not benefit livestock productivity, or more to the point, the livestock themselves (Telenged 1996). With the privatisation of livestock in 1990, all breeding programmes collapsed and herders reverted back to previous techniques without the use of artificial insemination and ceased cross-breeding with Russian livestock. These altered "super breeds" required extra fodder that was no longer available with the cessation of Russian support. A report on the state of Mongolia's genetic resources concluded that "the demand for a livestock breeding programme for improving quality and productivity was reduced for social reasons and the livestock genetic resources pool protection and improvement actions became stagnant. . . . Results from assessment of vet and breeding services shows that only vet services were effective" (Bynie 2002: 23). Herders reverted to relying on the animals' own mating behaviour to reproduce offspring, instead of artificial insemination, and found that the animals that were indigenous to Mongolia were better adapted to coping with the extreme local environmental conditions.

With the collapse of the USSR, Mongolians made a swift transition from a socialist to a market economy. This change was documented by Melvyn Goldstein and Cynthia Beall (1994), who found that although the herders were initially cautious about the change from a collective form of pastoralism, within a couple of years most had privatised their herds. Many Mongolians moved into the countryside to take up a herding way of life. Initially life became harder for Mongolian herders, as the Mongolian government had been reliant on aid from and trade with the USSR but found itself suddenly without financial support. During the socialist era, large factories converted wool into felt and milk into dairy products, but with the closure of most factories, the herders reverted to self-sufficiency, making the felt and dairy products by hand. Nowadays nomadic herders in Mongolia have reverted to traditional animal husbandry and veterinary techniques, as they are free to make their own decisions about their herds, rejecting the failed cross-bred varieties for the hardier and better adapted Mongolian livestock.

MONGOLIAN ATTITUDES TOWARDS HERD ANIMALS

A crucial aspect in understanding what the herd animal means to Mongolians is to engage with the herders' attitude towards the animals. This can only be accomplished through an analysis of Mongolian herders' world view and cosmology. In Chapter 2 I discuss how Mongolian herders' beliefs regarding a living animal are aligned with animist beliefs that occur in Central and Northern Asia (Harvey 2006a; Humphrey with Onon 1996; Vitebsky 2005). Some of the ethnographic and historical literature in Mongolia has investigated religion, specifically the Mongolian version of Buddhism introduced from Tibet, and a more ancient form of shamanism (Atwood 1996; Bira 2004; Bold 2001; Even 1991; Jagchid 2003; Pederson 2001; Siklos 1991). Although Mongolian herders in the Khangai mountain–steppe region could be categorised as conducting Buddhist or shamanistic practices, herders do not subscribe exclusively to one or other as a religion but have adopted a complex mix of rituals and traditions established over time. I describe the herding attitude as animist but this world view can coexist with practices and beliefs from Buddhism to shamanism, capitalism, or communism.

A shaman has the power to communicate with non-humans, particularly those beings that are part of the spirit world. The use of a shaman as an interpreter of the realm of spirits has an ancient past of more than 3,000 years, with evidence provided by deer stones (*bugan chuluu*) depicting inscriptions of shamanic flights to the sun (see Chapter 7). Siberian ethnography describes how the shaman can metamorphose, crossing from human form into another being, with an ability to adopt the external appearance of a bear, a crow, or even an inanimate object, such as a rock (Balzar 1996; Ingold 1986a; Potapov 1999). Shamanism is not a religious practice in the sense that Buddhism is. There are specific people who are shamans, and although the general population may go to these shamans for consultation they do not necessarily subscribe to shamanism as a concept. Prior to the Soviet era in Mongolia, because the shaman had the ability to negotiate with spirits, he or she was often employed to cure the sick, essentially functioning as the local doctor. As Urgunge Onon described to Caroline Humphrey, "The shaman is just like your GP (general practitioner, or family doctor). You go to him if you are ill" (Humphrey with Onon, 1996: 49).

The introduction of Buddhism meant that many of the rituals performed by shamans were either overtaken by, or incorporated within, Buddhist ceremonies. An example of this are rock cairns, or *övöö*, which are found on many of the prominent mountaintops or beside lakes in Mongolia. These sacred cairns can in fact be found in many pastoralist cultures throughout Eurasia. In Tibet the cairns often have yak horns placed on them; where sheep are important, there are sheep horns, whereas in Mongolia herders place the skulls of stallions. In Mongolia, these *övöö* were initially sites where herders would make clan sacrificial offerings of animals to the spirit world, officiated by a local shaman to request such things as rain or herd fertility (see Buxton 1926; Humphrey with Onon 1992). The *övöö* is considered to be the symbol of connections of group members sharing the same land. Each *övöö* on a prominent mountaintop represents the herding community (or *nutag*) within an area, such as a river valley. These same sites were diverted for use in Lamaist *övöö* ceremonies. Buddhist monks still conduct such rituals, particularly over the short Mongolian summer. Local herders make offerings, such as milk products, while monks read from Tibetan sutra. These *övöö* ceremonies incorporate festivities, specifically horseracing and wrestling (see Chapter 7 for a detailed description of horse racing). Atwood (1996) argues that ceremonies, such as the *övöö* ritual, are not shamanist at their core with a Buddhist veneer, as proposed by previous scholars; rather these traditions are a complex mix of both Lamaist and shamanist traditions.

The adoption of Tibetan Buddhism (sometimes called Lamaism) within Mongolia is not surprising, as the high-altitude Tibetan Plateau and Mongolian steppe have similar forms of nomadic pastoralism (Jagchid 2003). Early versions of Lamaism were introduced by Tibetan monks under the rule of Khubilai Khaan, in the thirteenth century A.D., but the penetration of Buddhism into herding society did not occur until the end of the sixteenth century. Mongolia adheres to this later form of the Yellow Hat (or Gelugpa) sect of Buddhism. Under the socialist regime, most of the established Buddhist monasteries were destroyed and monks were forcibly removed, resulting in the decline of Buddhism amongst the general populace (see Baabar 2004; Becker 1992). Throughout my field work in the Khangai region, I did not encounter any practicing local shamans. Most of those who existed in the past would have died out or ceased practicing during the Soviet era, when shamanic rituals were forbidden.[5] There has, however, been a resurgence in Buddhist practice since 1991, whereby Mongolian herders frequently ask advice from practicing monks, asking them to read from specific texts for such matters as increasing the number of livestock or curing a family member from a troubling illness (see Chapter 10). Knowledge of medicinal plants and the use of "magic medicine" (*em dom*) techniques to treat illnesses in both people and animals are likely to stem from ancient Buddhist and shamanic knowledge (see Chapter 9 for the use of so-called magic medicine and medicinal plants).

Recent literature regarding hunter–gatherer cosmology refers to shamans within an animist system but does not refer to shamanism as a concept (Harvey 2006a; Ingold 2000). Instead of categorising the herders' spiritual practices as either Buddhist or shamanist, I prefer to use the more encompassing term of "animism." The reason for this is that herders have seldom received the years of training required to be a monk or a shaman, but they ascribe to specific rules and traditions with regard to their herd animals that would be difficult to trace back as being derived specifically from Buddhism or shamanism. In using the term "animism," I am not describing the religious belief of Mongolian herders but their general cosmology, world view, and attitude towards other-than-human beings.

The original use of the term "animism", by E. B. Tylor, was to describe the belief that non-human entities harbour souls. He was using this form of belief in opposition to the Christian view, which considers only humans to contain a soul (see Harvey 2006a, for a summary on this earlier view of animism). Animism has been revisited more recently by scholars, largely

[5] Humphrey (1980) comments that there were few shamans in Central Mongolia (which includes the Khangai region) even by the beginning of the twentieth century.

conducting ethnography on hunter–gatherer societies (Bird-David 1990, 1999; Campbell 1995; Descola 1992, 1994, 1996; Viveiros de Castro 1998). The initial use of the term "animism" has become outdated because many hunter–gatherer societies do not believe in the equivalent of a "soul," as the soul is essentially a Christian concept. Animists are now thought to speak of spirits, something with heightened power, or as Humphrey with Onon (1996) describe it, as an "indeterminate energy" within other-than-human beings. One of the key ethnographies that inspired this interest in a new theory of animism is the work of Irving Hallowell (1976) on the Northern Ojibwa people. For the Ojibwa, even what we would class as inanimate objects, such as particular, individual stones, may be categorised as being alive or animate. He provides an anecdote of how an old man informed him that some stones are animate, yet most are not. Hallowell asked, "Are all the stones we see about us here alive?" The old man reflected for some time, then replied "No! But some are." Hallowell uses the phrase "other-than-human persons" to refer to those beings that are animate. Mongolians also believe that specific rock outcrops, or mountains that are distinct and of an unusual shape, or a lone tree, have heightened energy or power.

The word "animism" appropriately conjures up connotations of air, of breath, of life. Other beings are ascribed vigour and liveliness and thus are animated. The word for animal in Mongolian is *am'tan*. This is derived from the word for mouth (*am*), similar to the word for alive (*am'd*), and to the word for breath (*am'sgal*), which indicates that any being with a mouth, who breathes, is an animal. Mongolians evidently categorise people *with* other animals, as the Mongolian word for people or human is *khünam*. According to Even (1991), added to these notions of being alive and having an ability to breathe is the Mongolian concept of something akin to a "soul," which is a reflection of an animal's bones because they are an "image or shadow of the body" (p. 185). In this recognition of other beings as animated persons, Mongolian herders may in my view be considered as animists.[6] Herders express a form of animism that Graham Harvey (2006a: xi) describes as follows: "animists are people who recognise that the world is full of persons, only some of whom are human, and that life is always lived in relationship to others." Harvey says animists "inculcate locally meaningful ways of communicating with other-than-human persons, especially in order to express respect" (Harvey 2006a: 9).

[6] In discussing the concept of Mongolian herders' views of animals as "persons," I am not drawing on the anthropological literature on the self or on personhood, as this is a separate theoretical discussion (Eriksen 2001; Mauss 1938/1985).

CONCLUSION

Living with Herds is essentially an etho-ethnography about Mongolian lives in the Khangai Mountains, including both human and other animals, specifically ungulates. This book engages with the co-domestic relationship as an everyday and developmental process but is not an attempt to trace the origins, the exact point at which domestication all began in Mongolia. Instead, the emphasis is upon the significance of the human–ungulate relationship, between co-domestic herders and herd animals in the landscape of the Khangai Mountains. Here I reiterate my definition of a co-domestic relationship: the social adaptation of animals in association with human beings by the means of *mutual* cross-species interaction and social engagement. Mongolian herders identify the herd animals as a part of their lives and as indigenous to the mountain–steppe habitat. In other words, the horses, cattle, sheep, and goat herds all belong in the Mongolian landscape, as "beings in the world." I argue that Mongolian herders' attitude towards their herd animals is based on an animist perspective of the world, in which other beings are considered as persons and are expected to be accorded with respect. In this book, the herd animals themselves play a central role, as they both influence and are influenced by the herders. Through ethnographic examples, throughout the chapters to follow, I aim to demonstrate how the social relationships between herders and different species of herd animals are shaped and constructed.

2

A MONGOLIAN ETHO-ETHNOGRAPHY

INTRODUCTION

Throughout my descriptions in the etho-ethnographic chapters that follow, I aim to show that Mongolian herders view their herd animals as agents who participate in a reciprocal, co-domestic relationship. This level of coexistence requires a relationship that is based on trust and respect, as the herd animals are contained to a minimal extent yet choose to associate with the herding family of their own accord. The reciprocity between herder and herd animal works because they live in close proximity to one another within a core area, the "co-domestic sphere." My approach towards the human–other animal relationship is similar to Tim Ingold's in that we both aim to cross preconceived disciplinary boundaries between the natural sciences and the humanities and to take care to avoid the traditional dichotomy of nature versus culture. I do differ, however, from Ingold's conclusions with reference to a frequently cited paper on human–other animal engagement and domestication entitled "From Trust to Domination: An Alternative History of Human–Animal Relations" (1994). Within this chapter I discuss my contrasting view based on observations from the field with regard to Mongolian herders as pastoralists and their attitudes towards their herd animals.

There has been a tendency in literature on nomadic pastoralism to unify all peoples together under a common framework (Barfield 1993; Dyson-Hudson & Dyson-Hudson 1980; Ingold 1994; Khazanov 1994). Instead of discussing the animals as property, commodities, or as metaphors for human behaviour, my focus is on the herd animals as actors in their own right. My observations in Mongolia show that differences in spiritual beliefs and practices provide a good indication that attitudes are variable and that these differences are indicative of the relationships between the people and the animals they associate with. In this chapter, I emphasise how Mongolian pastoralists' beliefs and ways of behaving towards animals differ from pastoral attitudes in the Mediterranean but are similar to other ethnographic accounts of animist hunter–gatherer practices, particularly in Siberia.

Significant Others

In a paper in 1969, Burton Benedict appealed for more integrative research across species. He said, "A major task for anthropologists and ethologists at present is to develop concepts which can be applied cross-specifically, so that we can compare social behaviour of various animal species including man. . . . Social anthropologists frequently use *caveats* against ethnocentricity, yet their studies are invariably anthropocentric" (Benedict 1969). However, the literature since Benedict made this statement has often remained disparate between the two disciplines of social anthropology and ethology, as researchers tend to focus on either people or other organisms with a tendency to make them mutually exclusive areas of study. In a review of the anthropological literature on domesticated animals, E. Shanklin (1985: 380) states that "I believe that the investigation of human and animal interaction may well be one of the most fruitful endeavours of anthropology."

Since I started my research in 2004, there has been a veritable explosion of cross-disciplinary work on human–animal studies, albeit focusing predominantly on the household pet or western farm animal. Rebecca Cassidy (2007: 4) states that "by including only the kinds of domesticatory relationships that are embodied by selective breeding and other features of post industrial agriculture, we exclude 97 percent of the estimated 80 billion people who have lived." *Living with Herds* focuses on human–animal relations (or more precisely, human–other animal relations) from a cross-cultural perspective. These relations can be approached from many angles, but I have chosen two obvious perspectives, those of the other animal (ethology) and those of the other human (ethnography).

Dominique Lestel and colleagues (2006) have proposed exciting new theoretical terrain, outlined in their article entitled "Etho-Ethnography and Ethno-Ethology." Ethno-ethology "grants all living beings the status of relational beings, that is, agents interacting on the phenomenon of 'culture' that was hitherto reserved for human beings" (p. 168). Etho-ethnology "seeks to describe and understand how humans and animals live together in hybrid communities sharing meaning, interests and affects, articulated around jointly negotiated significations" (p. 173). Lestel (2006) challenges academics from the fields of ethology and ethnology to investigate new fields of enquiry, studying the hybrid communities comprising both humans and other animals, in order to interpret their shared lives and socially complex interspecific associations. Barbara Noske (1993) states that anthropologists should consider animals' relationships with people just as they consider humans' relationships with animals, rather than persisting with an anthropocentric point of view. According to Harvey (2006b), ethologists who now recognise the cognitive competence of animals could gain considerably from a discourse with animists, such as Mongolian herders, to learn more about the perspectives of animals. He argues that we can climb out of the Cartesian gulf we have dug: "Gaining a relational and participatory perspective on the job of academia may greatly enhance the effectiveness of our academic work" (Harvey 2006b: 15). Eduardo Kohn (2007) makes steps towards developing an "Anthropology of Life" that is not just confined to the human but our entanglements with other living selves. He refers to this kind of anthropology as "transspecies ethnography" (p. 18).

My work aligns with that of Tim Ingold's in many respects as he is also concerned with closing the gap between the humanities and the natural sciences and has a general concern for human–animal relations, including conceptualisations of nature and culture. Ingold (1996) distinguishes between two kinds of nature: "really natural" nature, studied by natural scientists, and "culturally perceived" nature, studied by social anthropologists. Ingold proposes that this division of academic labour between the broader framework of natural science and the humanities is associated with the separation of humanity from the world of nature. He summarises his view in the following way (Ingold 1990: 224):

> Social anthropology has explicitly defined the person as its object of study, but only by setting itself up in opposition to a biology of organisms, thereby driving a wedge into the human being, splitting it irrevocably into two mutually exclusive parts – the one individual, the other social. The result has been to perpetuate a separation between humanity and nature that has had fateful consequences in the history of western civilisation. The most urgent task for contemporary anthropology is to overcome this separation, and to re-embed the human subject within the continuum of organic life.

Throughout his influential book, *The Perception of the Environment*, Ingold (2000) emphasises that both people and other animals are beings in the world. They do not just exist in the mind but are part of the landscape. One of the means of being in the world is to adopt the social anthropological tool of participant observation, in which the researcher lives with and becomes involved in the subjects' daily lives by being an active participant in society (in this case by communicating and interacting with both herders and herd animals). This anthropological perspective allows for a synchronic approach to herder–herd animal relations within a specific social and cultural context, the Khangai mountain steppe of Mongolia.

Jacob von Uexküll was a founder of ethology and was an original proponent of the biosemiotic approach. He was one of the first to describe an animal's "subjective universe" by the German term of *Ümwelt*, or the way an animal perceives the world, encompassing this view in a "theory of meaning" (Uexküll 1957). A spider's web is a physical extension of the spider itself but it is also an exploitation of the perception of a fly, as it fits the subjective universe of the fly so well that the spider can literally capture it (see Jackson 1992). An acute awareness of another being essentially involves a perception of that being's *Ümwelt*. Within this book I have attempted to engage in a multispecies herding universe, the Mongolian herding *Ümwelt*.

The herders I met were not particularly verbose when communicating with one another, or towards the animals, but they engaged in a physical sense. Children were not taught through verbal lecturing and reasoning; they learnt through a gradual accumulation of embodied knowledge by actually participating in an activity, through what Michael Jackson (1983: 337) refers to as "kinaesthetic learning" (a different kind of learning from schooling within a Western education system). This was necessarily the way I learnt during my time in Mongolia, essentially employing the main fieldwork method in anthropology – participant observation. I was limited by the extent of my understanding of Mongolian (and ungulate) as a language, so I was forced to learn in a more embedded way through living in the environment the herders lived in, engaging in the same tasks the herders engaged in, and through actively interacting with both herders and herd animals.

Moving, visual images are particularly relevant for conveying interspecies relationships, as there is less engagement by the means of language and dialogue and a greater reliance on body language and sound to convey intention. I shot footage that incorporates co-domestic humans and other animals as the subject material, employing predominantly ethnographic filmmaking principles but using some elements of wildlife filmmaking techniques to aid in our understanding of "significant others" (for more details on my approach to filming in the field, see Fijn 2007).

There is scope to make an original contribution to our knowledge by presenting a process of enquiry, and this is what I set out to achieve with my form of observational filmmaking (see MacDougall 1992, 1998, 2006). Unlike other media, film is not reliant on words to convey understanding and meaning but can effectively show emotion and body language as means of communication. One of the great aspects of the use of visual material is that it can be used to break new ground across unexplored territories (see Banks 2001). As David MacDougall (2006: 220) writes, "visual anthropology may offer different ways of understanding, but also different things to understand."

PERSPECTIVES ON THE DOMESTIC ANIMAL

In a wide-ranging review of the literature on human–animal relationships, Molly Mullin (1999) recognises that, since the late 1980s, studies on pastoralism have focused mainly on political and economic change and the pastoralists' dealings with government and development organisations. There has also been a focus on how pastoral nomads use, relate to, and exist within specific environmental conditions (Barfield 1993; Feeny et al. 1990; Fratkin 1997; Hardin 1971). In ecological anthropology, herd animals are treated largely as a means for human sustenance (Harris 1992; Heston 1971), or as interactive factors between humans and the environment (see Leeds & Vayda 1965; Shanklin 1985). However, in their review of the nomadic pastoralism literature, Dyson-Hudson and Dyson-Hudson (1980) recognise a lack of emphasis on the interactions between nomads and herd animals themselves. Herd animals are generally described in a biological manner but they are not considered as a key part of pastoral nomadic societies. There has been considerable focus on nomadic pastoralism in relation to the environment in Inner Asia (e.g., Dunlop 2000; Humphrey & Sneath 1996a, 1996b; Neupert 1999; Sneath 2000), but less regarding the social importance of the herd animals. I have intentionally focused my attention towards the significance of the herd animals as entities.

In her review of the socio-cultural literature on human–animal relationships, Mullin points out that research in the humanities often approaches human–animal factors to examine aspects of human society, rather than the human–animal relationship being of particular interest of itself. The exception to this is in some research based in Africa, where the herd animals, particularly cattle, are recognised as an inherent part of herding society. Mullin (1999) suggests that this different approach exemplified in this African research engages in a more celebratory fashion with pastoralism as a productive and viable way of life. Galaty (1989), for instance, explores the value of cattle in terms of how Maasai classify them according to individual characteristics and genealogy.

This focus on the significance of cattle was instigated by E. E. Evans-Pritchard (1940, 1956) in his groundbreaking ethnography. He describes, at one point in *The Nuer* (1940), how he was frustrated in the field when Nuer conversation constantly reverted back to talk of cattle. Fortunately for me, the Mongolian herders' tendency to talk about the herd animals was relevant to my own research and I welcomed any conversation regarding herd animals. Animals are an integral part of the colloquial conversation of Mongolian herders; for example, "*mal sureg targan tavtai yu?*" is a greeting, which can be literally translated as "are the animals fat and comfortable?" When discussing East African pastoral societies, Fukui points out that "as they live their whole lives with their domesticated animals, the animals become an indispensable focal point not only for human relations within society but even their entire world view" (Fukui 1996: 320). Through the discussion in subsequent chapters, my intention is to show that herd animals are integral to the world view of Mongolian herders (see Chapter 10 in particular).

INGOLD'S PERCEPTION OF THE ENVIRONMENT

Anthropologists writing about hunter–gatherer societies have given considerable attention to human–animal relations, particularly with reference to so-called wild animals. In the hunter–gatherer literature there are debates such as the dichotomy between nature and culture, the differences between animist and totemic systems, and the personhood of other-than-human beings. Tim Ingold has contributed considerably to these debates. His writing (particularly Ingold 2000) has proven important for many of the ideas discussed in this book.

Ingold's ethnography initially focused on Saami reindeer herders and he continues to focus his writing on peoples from the northern circumpolar region (Ingold 1974, 1980, 1986a, 1986b, 1994, 1996, 2000).[1] His early work on human interactions with reindeer primarily focuses on the economic outcomes of pastoralism for humans, referring to human–reindeer social relations as an "exploitation" of the reindeer for economic gain (Ingold 1974, 1980; also see Aikio 1989 for an opposing view). This perspective aligns with Marx's theory of value, in which the raw materials of nature are "primitive instruments of labour" (Marx 1867/1976). Both Marx and Engels describe production as the essential criterion that sets humans apart from all other animals. In his more recent writing, Ingold recognises that he had an anthropocentric, economic bias in his earlier work, which he attributes to the more general human versus nature dichotomy still evident in Western society (Ingold 1994).

In recent decades, anthropology literature has emphasised the need to avoid imposing nature–culture dichotomies upon other cultures, as many societies do not conceptualise the world in this way (for an ethnographic example, see Strathern 1980). Descola succinctly summarises the views of anthropologists and historians, who say that "conceptions of nature are

[1] Reindeer are the main herbivores in the northern regions of Eurasia. Many ethnographies in this region emphasise the strong relationship of hunters and pastoralists with reindeer (for examples see Aikio 1989; D. Anderson 2000; M. Anderson 1986; Carruthers 1913; Khazanov 1994; King 2003; Mirov 1945; Vitebsky 2005).

socially constructed, that they vary according to cultural and historical determinations, and that, therefore, our own dualistic view of the universe should not be projected as an ontological paradigm onto the many cultures where it does not apply" (Descola 1996: 82).

Mongolian vocabulary does contain different words for "natural" (*baigal*) and "cultural" (*soyol*) but the concepts are not distinctly dichotomous, opposing one another. The word for culture, *soyol*, comes from the noun *soyo*, or "sprout." Accordingly, "culture" can be seen as "an offshoot from the parent plant (one 'existence', or nature), and, becoming itself a parent plant it gives rise to further offshoots (in succeeding 'existences')" (Tseren 1996: 148). The word for nature (*baigal*) is closely related to a "state of being" (*baidal*; see Humphrey & Sneath 1999). It is clear that, in terms of the Mongolian conception of the world, nature and culture do not correspond in the same way as the Western binary of nature versus culture.

Val Plumwood explains that the Western human–nature dualism destroys the bridge between the human and non-human, and "as the essentially human is disembodied, disembedded and discontinuous from the rest of nature, so nature and animals are seen as mindless bodies, excluded from the realms of ethics and culture" (Plumwood 2003, p. 3). In Western tradition there is a tendency to perceive humans as belonging to a totally different order, expressed through "culture," whereas other beings are in a totally different realm, "nature." In reference to an alternative to this Western social construction of nature, Bruno Latour writes that "all nature-cultures are similar in that they simultaneously construct humans, divinities and non-humans…none of them inhabits a world of signs or symbols arbitrarily imposed on an external Nature known to us alone" (Latour 1993: 106). Note that Latour (1993) refers to the existence of "nature-cultures," or, as in Donna Haraway's terminology, the two words become one, as "naturecultures" (Haraway, 2003, 2007).

Ingold (1994) proposes that we revise the orthodox distinctions between what we term as "wild" and "domesticated." He argues that the separation of culture from nature, and of domestic from wild, derives from a world view that was perpetuated by Christian theologians, who argued that man was created in the image of God and was inherently superior to all other living creatures because man possessed a soul. We presuppose that, through the domesticating of animals, we have somehow intervened and transcended beyond the natural world into a cultural one. Instead of considering domestication as a great feat of engineering, in which the domesticated animal is "made" as an object, Ingold (2000) suggests that we think of the animal as being "grown" or nurtured by humans, just as children are "grown up." He proposes (Ingold 2000: 76) an alternative to the monotheistic, dichotomous view, arguing that both people and other animals are beings in the world:

> I am suggesting that if we rewrite the history of human–animal relations, taking this condition of active engagement, of being-in-the-world, as our starting point. We might speak of it as a history of human concern with animals, in so far as this notion conveys a caring, attentive regard, a "being with." And I am suggesting that those who are "with" animals in their day-to-day lives, most notably hunters and herdsmen, can offer us some of the best possible indications of how we might proceed.

This is one reason why I have given this book the title of "Living *with* Herds," as it implies a state of "being with" the herd animals, not just viewing them as a means of production or purely for economic profit. Vinciane Despret also emphasises the importance of "with-ness," an articulation of being with someone. She refers to the ethologist Konrad Lorenz's experiments with jackdaws as being as much a "jackdaw-with-human" as "human-with-jackdaw" relationship (Despret 2004). This day-to-day interaction with other species is a key reason why I was interested in living with Mongolian herders and their herd animals, as herders have an intimate knowledge of herd animals through the process of *growing up* amongst them from birth.

TRUST AND DOMINATION

In Mongolia, the *seter* animal is easily recognisable within a herd, as it has a blue silk sash (*khadag*) tied around its neck. This animal is a sacred or holy animal, who acts as a spiritual protector (*malyn ezen*). A monk blesses this individual ungulate and instructs the herder not to kill, exchange, or give away the animal (see Chapter 10). Eventually it will die of natural causes. In effect, a *seterlekh* ceremony performed by a local lama is an antisacrifice. One animal is protected to ensure the health and safety of not only other members of the herd but the herding family and even the surrounding landscape. This sacred animal is symbolic of the herders' relationship with the herd animals as a whole.

In his well-known article in the domestication literature, "From Trust to Domination", Tim Ingold (1994) argues that hunter–gatherers form a relationship of trust between the hunters and their animal prey and that, in contrast, pastoral peoples have transformed this relationship of trust to one of domination with their domesticated animals. In other words, it is the domestication of animals that Ingold distinguishes as the point at which a divide in human–animal relations formed. In terms of hunters, he refers mainly to northern peoples, such as the North American Cree, and their view that "consumption follows killing, as birth follows intercourse" in a constant cycle of regeneration (Ingold 1994: 67). Counter to Ingold's contentions that pastoralists have a dominating view regarding their domesticated animals, I have found that Mongolians have much the same attitude as the northern hunter–gatherers to which Ingold refers in that the relationship is viewed as a reciprocal one.

To avoid confusion about what is meant by "dominance," I want to emphasise that Mongolian herders do behave in a dominant manner towards their herd animals to retain control over the herd. As I discuss later in the etho-ethnography, they take the place of the lead animal within the hierarchical structure of the herd (see Chapter 6). Inherent in this dominant role, however, is a need for a high degree of *trust* from the entire herd, otherwise the herd animals would not accept the role of the herder and could readily flee, as there are no fences to obstruct them. This *behavioural* domination, which would be better described as leadership, is quite different from the "trust to domination" that Ingold is discussing. Ingold's "trust" refers to the point in time when an animal is killed, that moment when a hunter comes face to face with a deer. My point is that the Mongolian herders' attitude towards killing their herd animals is *not* markedly different from the attitudes of northern hunting peoples, as both believe that the essence of the being will pass on from one life form into another upon death and an inherent part of the death of an animal is respect. Furthermore, trust is an essential element of herder–herd animal mutual co-dependence.

Ingold's stance is clear when he states that "like dependants in the household of a patriarch, their status is that of jural minors, subject to the authority of their human master. ... In short, the relationship of pastoral care, quite unlike that of the hunter towards animals, is founded not on a principle of trust but of domination" (Ingold 1994: 16). I propose here that Mongolians do not have an attitude of human mastery over their herd animals, in which the animals are treated as serfs. Tani (1989, 1996) relates how the castration of male sheep and goats and their relationship to the herder was originally derived from the attitude of an ancient Sumerian master to his serf, or the eunuch. Tani's field area was based in the Mediterranean and Middle East. Pastoralists in this area believe in Creation, in the Old Testament, and that God created man and entrusted all animals under his control. Tani contrasts this biblical attitude towards animals in the Mediterranean and Middle East with Mongolian pastoralists who, in having a *seter* (or sacred animal), are more aligned to the belief systems of Siberian hunters. I agree with Tani's conclusions, even though Tani's work was a likely influence for Ingold's comparison of domesticated animals with serfs (as Ingold's 1994 article references Tani's 1989 paper).

Importantly, part of Mongolian herders' subsistence still relies on occasionally hunting animals and gathering medicinal plants. As Bold (2001) points out, pre-Mongolian and Mongolian peoples have combined hunting with livestock keeping for thousands of years. Even (1991) states that, in terms of shamanism, the Mongolian relation with Siberian hunters should not be overlooked, as similar alliances were in practice with the hunter and the shaman in their quest for game and the spirit world. Siberian hunters believe that there is a "chief" animal, or master of the forest, such as a deer, which may be especially large or beautiful, and this individual is not to be killed by hunters as a way to ensure the protection and future reproduction of other animals within that species. Mongolian hunters also ascribe to *Manakhan*, the provider of game and prosperity, which according to Even (1991) and Bawden (1968) is similar to the Siberian concept of the master of the forest. Tani aligns Siberian hunters with Mongolian herders in that *both* "justify their killing and consumption of animals by prohibiting the slaughter of a special individual held to be responsible for the herds' prosperity" (Tani 1996: 411). In both my own and Tani's view, the Mongolian herders' view regarding the *seter*, or sacred animal, is functionally equivalent to that of the master, or chief animal of the forest amongst Mongolian and Siberian hunters.

Even though Tani specifically separates the views of pastoralists from the Mediterranean and the views of pastoralists in Mongolia, Ingold does not do this and extends his view of domination to all pastoral peoples. In Ingold's (1986a) view, hunter–gatherers attribute animal masters with governance over wild animals, whereas pastoralists are human masters over their domesticated animals. In Tani's view, pastoralism in the Mediterranean region is one of human dominance over a subjugated flock and is in accordance with a vertical, hierarchical structure described in the Old Testament. However, according to Mongolian herders, the *seter* animal and other sacred beings (such as specific mountains, trees, or the weather) hold both human and animal well-being in their power. In Mongolian pastoralism, humans are not viewed as all powerful and have not taken the place of the animal chiefs or masters in hunter–gatherer societies, as the *seter* has taken this role. Mongolian herding cosmology is based on an animist perspective, which has a more horizontal, reciprocal structure in which humans can reincarnate into animals, specific sacred animals have increased power, and herder–herd animal relationships are based on reciprocity.

Piers Vitebsky (2005) writes that Eveny reindeer herders in Siberia understand the interrelationship between hunter and prey in terms of the animal's *giving* itself to the hunter if he treats its body and soul correctly. Dreams are symbolically important to both Eveny reindeer herders and Mongolian herders, as dreams can predict significant events in a person's life.

An example of this occurred when I returned unannounced to Choijo and Naraa's *ger* after being gone for more than a year. Naraa was overwhelmed with emotion but not surprised to see me. She told me that the previous night she had dreamt that I would come back to Mongolia to visit them and that she had spoken to her grandsons that morning about her dream, foretelling my arrival that day. Eveny reindeer herders believe that a dying animal in a dream is a substitute for a person who is living at the animal's expense, such as a collapsed reindeer representing someone's terminally ill father in a dream (Vitebsky 2005). This interchangeability of life between an animal and a human in the Eveny's interpretation of dreams may be similar to the Mongolian perception that the "giving" of life to a *seter* animal balances the death or illness of a family member.

It is well known that one of the key ideas of Buddhism is one of interconnectedness, inherent in the concepts of *kamma* and rebirth (McDermott 1989). Both hunters and Mongolian pastoralists believe that the animal must be killed according to quite specific rules, and that bad or harmful consequences will ensue if the killing is not done with a high degree of respect (see Chapter 10). Ingold argues that domesticated animals are presumed by pastoralists to lack the capacity to reciprocate but it is clear that Mongolians engage in a reciprocal relationship with their herd animals and that their cosmology is representative of this reciprocal relationship (see also Kaczensky 2007).

The herders' selection of a *seter* animal is a key to their attitude towards their co-domestic animals as a whole. The selection of sacred animals from amongst others of the same kind is also reflected north of Mongolia, amongst the reindeer herders of Siberia, and to the south of Mongolia amongst the nomadic herders of Tibet. Ben Campbell recognises that the Tamang of Nepal also attribute special characteristics to a single animal. He refers to the Tibetan word *tsehtar*, the equivalent of the Mongolian *seter*, defining it as giving the animal "life-freedom," liberating the animal (Campbell 2005: 95). James Serpell (2000, 2005), in agreement with Ingold (1994), emphasises the differences between the hunter–gatherer and herder systems of belief. He argues that animals were killed by herders throughout ancient civilisations as sacrificial gifts to the gods, a means of solving the moral dilemma of killing an animal. Yet, again, Serpell's statements about herders cannot be generalised to all pastoral herders because Mongolians do not offer their animals as sacrificial gifts to the gods; instead, they offer a sacred animal, essentially as an antisacrifice, whereby they *refrain* from killing an animal.

Campbell finds Ingold's paper overly generalizing in saying that pastoralists are dominating: "There are sufficient reasons to question Ingold's assertion of a generic dominatory logic in pastoralism as compared to hunter-gatherers" (Campbell 2005: 96). John Knight also critiques Ingold's article, pointing out that "hunters necessarily lack familiarity with the individual animals they hunt, even if, as they gain experience over time, they acquire a generic familiarity with the patterns of behaviour associated with the kind of animal in question. By contrast, domestication does provide the temporal and spatial conditions for human-animal intimacy to emerge" (Knight 2005: 5). It is evident that peoples living according to animist traditions have similar concepts of sacred animals, who act as protectors and representatives, whether it is for the herd, for their species, or for all living things. These belief practices are quite different from monotheistic beliefs that are based on the Old Testament.

In summary, it is my view that the cosmology of the arctic hunter and the Mongolian herder are similar in relation to their attitudes towards the life and death of specific beings. A herd animal's life is interchangeable with a herder's life, and vice versa. Through the etho-ethnography that follows, I aim to show that instead of a dichotomous division between human and non-human animal, herders include themselves and herd animals in a constant struggle for balance in the cycle of life and death. This perfectly aligns with Ingold's comments in relation to animist systems in northern arctic hunters. He says that "the animals offer something of their potentiality and substance to human beings so that the latter may live, while humans, in return, through the proper treatment of the animals in death, ensure the release of their life force and hence their subsequent reincarnation" (Ingold 2000: 114).

CONCLUSION

Mongolian herders have a world view and cosmology that is reflected in and shaped by their relationship with others, particularly herd animals. "Animism" is often a term used in anthropology to describe the relationship of northern arctic peoples' relations with other animals and plants. I argue that Mongolian herders are animists in a similar sense, because herd animals are treated not as objects but as subjects, thus encouraging a relationship of mutual respect. Mongolian herders relate to herd animals as other persons who are capable of expressing feelings and emotion; each individual animal being is ascribed recognisable characteristics and a distinct personality (see Chapter 3). Mongolian herders view themselves as being in a reciprocal relationship with the animals they herd, such that if the herding family works hard to nurture and provide for the animals then in turn the herd animals will nurture and provide for the herding family, allowing them to live mutually interdependent, happy, and prosperous lives.

My work aims to cross many boundaries: Mongolian herder–herd animals, human–non-humans, nature–culture, ethology–ethnography, and textual anthropology–visual anthropology. This research is encompassed within the broad rubric of "anthropology," both biological and social, but bringing to this project a grounding in ethology. This involves an empirical, phenomenological approach, a being in the world, rather than a subscription to a purely observational, interview-, or survey-based analytical approach. As the contributors to the recent book *Animals in Person* (Knight 2005) aimed to do, I am engaging with herd animals as subjects rather than objects, the ungulates in Mongolia as a part of society rather than just symbols of Mongolian society, and herder relationships *with* herd animals rather than simply human representations of other animals.

SUMMER

Plate 9. Tibetan script on carved stone.

Plate 10. Collecting firewood with oxen.

Plate 11. Three jockeys resting between races.

Plate 12. Winning horse with medals.

Plate 13. Winning stallion with jockey and trainer.

Plate 14. Wrestling at a local *Naadam*.

Plate 15. Resting while out herding.

Plate 16. Deer stone (*Bugan chuluu*) lying out on the grassland.

PART TWO

THE SOCIAL HERD

3

SOCIAL SPHERES

Just as a human can live as a hunter-gatherer or in a prison cell, so a cow can live as a free-ranging grazer or in a factory farm. But neither the human nor the cow actually changes their inherent behavioural impulses or their innate powers of perception, although these may be suppressed by lack of opportunity. (Clutton-Brock 1994: 30)

INTRODUCTION

The domestic sphere of the encampment does not have sharp boundaries; people and animals move within and beyond this sphere frequently, but most of the herders' and herd animals' time and energy are spent within this co-domestic sphere. In the first part of this chapter I describe the importance of the social landscape of the herding encampment as the central sphere for establishing the co-domestic relationship between herding family and herd animals. Much of the herders' and herd animals' world is focused upon this core encampment, even though the location of the encampment itself migrates and changes with the seasons. The encampment is, therefore, not a fixed space that can be quantitatively measured; both people and animals inhabit it as an *ecosocial sphere*. Richard Nelson (1983: 243) describes a similar interweaving of life histories into the surrounding landscape in relation to the Koyukon people: "Each living individual is bound into this pattern of the land and people that extends throughout the terrain and far back across time."

Mongolian herders have built up a vast wealth of knowledge about the social behaviour of herd animals from interacting with them throughout their lives. Knowledge of animal behaviour and husbandry techniques in Mongolia is gained through oral tradition, through a young herder learning by trial and error, or from watching more experienced herders. The herders utilise this knowledge of herd behaviour to influence the social structure of the various herds. The second part of this chapter is divided according to the structure of the herds of horse, cattle, sheep, and goats. I begin each section with a description of the behaviour of feral or wild herds from similar genera in other parts of the world, and I compare this literature with my knowledge of the social structure of the herds in Mongolia. I emphasise the social behaviour of the different herds of horse, cattle, sheep, and goats and link this with the social behaviour of their human counterparts in an attempt to detail this "anthropo-zoo-genetic" relationship between herders and the various species of herd animal in the Khangai Mountains of Mongolia.

PART I. THE SOCIAL LANDSCAPE

The use of the word "environment," according to Graham Harvey (2006b: 12), "too often implies (to those of us used to modernity's consumerist worldview) a human-centred vision of resources to be exploited, with or without respect." I am using the word "landscape," which in Mongolian could be translated as *baigal' gazar*. My use of the term "landscape" is in reference to Anna Tsing's use of the term as the "configuration of humans and nonhumans across a terrain" (Tsing 2004: 173) in relation to one another. I am deliberately avoiding the use of the term "environment" (*baigal' orchin*) because Mongolians adopted this concept only more recently, perhaps from a Russian translation of the term.

To Tim Ingold (1993, 2000), the term "landscape" does not comprise "land," nor is it analogous to "nature" or "space." Ingold states that "in a world constructed as nature, every object is a self-contained entity, interacting with others through some kind of external contact. But in a landscape, each component enfolds within its essence the totality of its relations with each and every other" (Ingold 2000: 191). He views the temporality of the landscape as a perspective on dwelling. Accordingly, "people do not import their ideas, plans and mental representations into the world, since that very world … is the homeland of their thoughts. Only because they already dwell therein can they think the thoughts they do" (Ingold 2000: 186) and, I would add, behave in the manner that they do. Earlier, in his seminal work, Gregory Bateson (1972) proposed that the mind is not limited by the body. Our mind is enmeshed within our "landscape"; rather than the concept of the mind as a separate entity, confined within a brain, Bateson instead proposed a phenomenological perspective of "being in the world."

The influence of the landscape upon Mongolian herders and species of herd animal cannot be neglected, particularly with regard to the domestic sphere or encampment (*khot ail*). An encampment is what Khazanov (1994) refers to as a "nuclear community," Evans-Pritchard (1940) as a "camp," and Barth (1961) as a "co-operative herding unit." Humphrey and Sneath (1999) refer to a "residential group," which comprises an extended family group with grandparents, parents, and grandchildren who share tasks and combine their animals together as one herd (in the mountain steppe a residential group usually consists of two to six separate *ger*). Some members may be unrelated, but this situation happens less frequently than that in which family members are related. According to Vreeland (1957), the Mongolian term *khot* represents a geographical entity, whereas *ail* represents the social unity of an encampment. Here, I am concerned with the ecosocial entity of an encampment (the *khot ail*), which is essentially the co-domestic sphere for herder and herd animal.

Perceptions of a Landscape

At first, I perceived the Mongolian countryside as a remote, wild, and potentially dangerous place. I was wary of falling from a horse, being kicked by a large animal, attacked by wolves, falling through ice into the river, or merely becoming ill from the change of diet. I felt anxious about how I would have access to medical care if a serious accident occurred. Leaving the winter site to get to an urban centre, both in the Arkhangai and Bulgan field sites, required riding a horse to a neighbouring encampment that had a Russian jeep (which might not be functional or contain any fuel), followed by at least two hours of rough driving to the nearest settlement (*süm*), and then another seven to twelve hours' drive from the provincial centres of Erdenet or Tsetserleg to the capital of Ulaanbaatar. I became more relaxed and used to such long distances to reliable medical facilities as I began to feel safe within the sociality and communal atmosphere of the herding encampment. I felt protected from wolves in the encampment and was reassured by the herding family that the wolves are afraid of humans, so they are unlikely to attack (see Chapter 9 for more details on wolves). I realised that as long as the dogs in the encampment recognised me as part of their domestic sphere, they trusted me and would protect me, just as they protected the extended family from prowling wolves or strangers intent on stealing livestock. The herd animals, too, provided me with a sense of belonging in their herd community. As I began to know and recognise individuals, the cows no longer attempted to move away when I approached during milking.

Mongolian herders do not perceive the landscape as a remote wilderness because it is their home and they are an inherent part of the ecosocial landscape. I asked Naraa to draw the main features of the landscape in the Jargalant Valley. Naraa drew the animal shelters and the round *ger* to

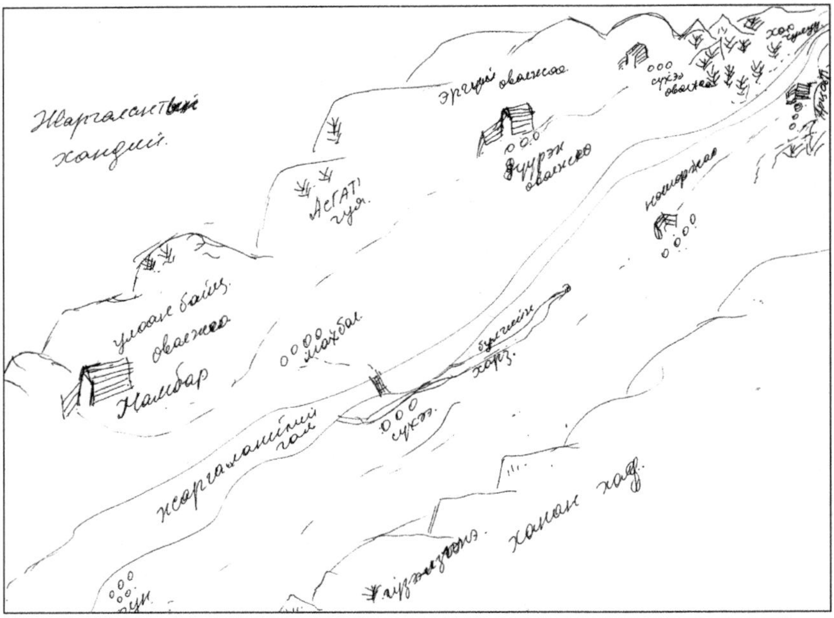

Figure 3.1. Naraa's drawing of the Jargalant Valley.

designate the seasonal encampments within the valley (see Figure 3.1). It is interesting that the encampments, bounded by the mountains and on either side of the river, are the prominent features of her landscape. Anderson (2000) refers to this interrelation between person and place amongst Evenki (related to the Eveny) reindeer herders as a "sentient ecology," whereas Kohn (2007) refers to a similar intersubjectivity between the Runa in Amazonia as an "ecology of selves." Both of these terms provide eloquent phrases for the Mongolian herders' relationship with their surroundings. What I had perceived in my mind as a wild, remote, and potentially dangerous environment soon became a landscape where I felt at home (see Jackson 1995, *At Home in the World*). My own perception of what was wild and what was domestic changed as my perspective of the landscape changed.

The times in Mongolia I enjoyed most were spent lying in the grass on a hillside, watching the sheep and goats graze, with the horse I was riding for the day cropping at the grass next to me. I valued the escape beyond the surrounds of the encampment. If I spent too much time within the confines of the encampment, I would begin to feel that my world had become too insular and restricted. The Mongolian herding encampment is a world unto itself, and the herding family within this encampment does not venture beyond this realm without a specific purpose and destination in mind. Even children tend not to play far beyond the encampment.

Although herders do not extensively explore the surrounding valleys on a regular basis, they have a good knowledge of surrounding topographical features (all of which are named).[1] They know where good pasture is located for the herd animals, where to find thin birch poles to chop up as firewood, where plants can be gathered during the summer to supplement the pastoral diet, and where to gather medicinal plants in autumn. Their home range, similar to the home range of the animals, decreases in the winter because of the extreme cold and the need to be in close proximity to shelter. In contrast, herders will go further afield in autumn to find good pasture for the animals or to collect the medicinal plants required for the following year. Herding the animals to distant pastures, or collecting plants, are ways in which women have the opportunity to relax from the work required around the *ger* and surrounding animal shelters. Beyond these activities, the encampment is the main place of existence, especially for women, the elderly, young children, dogs, and young herd animals (see Chapter 9, Figure 9.1).

Each herding family has occupied the campsites and nearby pasture over many years, and the valley as a whole has been grazed for hundreds, if not thousands, of years. This was reemphasised to me when Dogsomjav pointed out to me the location where Chinggis Khaan had rested on the edge of the Selenge River, just a few kilometres from her winter encampment. Again, when we moved to Dogsomjav's summer encampment, I noticed a stone tablet with a Tibetan inscription etched faintly onto the face, lying just near the sheep and goat pen. Such a rich history is deceptive when the landscape looks so sparsely inhabited to the outside observer's eye. The river valleys where the herders live are not "wilderness" at all but have been altered by herders and herd animals, becoming an inherent part of the surrounding landscape over the centuries (see Tsing 2004, for a similar perspective on secondary forest from swidden agriculture). All components of the ecosystem in Mongolia are intact, where humans, predatory wolves, the breeds of herd animal, and the grasses the livestock graze upon could all be categorised as indigenous to Mongolia. In the case of Mongolian pastoralism, it would be simplistic and inaccurate to draw a line between where "wild" begins and "domestic" ends (see Chapter 9).

[1] I do not mean to imply that herders are not knowledgeable about their surroundings and animals beyond those that are within the domestic sphere. Mongolians have many subcategories for describing the wild animals they hunt; for example, there are separate words for age and sex categories in deer: *buga* (deer, buck), *soroo* (doe), *maral* (hind), *zolu* (young female deer), *roodoi* (young fawn), *zorgol* (yearling), *comagai* (two-year-old deer), *tsökh* (three-year-old deer), *tair* (young male deer), and *turag buga* (mature male deer).

The Core of the Co-Domestic Sphere

The core of the co-domestic sphere is the family *ger*. In winter a Mongolian herding family relies on the *ger* for survival from the cold and it forms the centre of all social activities, including interactions with the young, newborn herd animals. Divangar[2] told me that herders think of the structure of the *ger* as analogous to the structure of an animal. There is an outer canvas layer, which represents the skin, while the insulating layer(s) of felt (*esgi*) is like the fat layer on an animal. In summer the felt is removed but there are as many as three separate layers of felt in winter, each approximately an inch thick. The inner lattice walls and poles encircling the roof are like the bones of the animal. This metaphor is also used by Siberian reindeer herders and is incorporated in local Eveny legend: "a reindeer rescues two little children from an evil spirit by turning its body into a living tent, with its ribs serving as the framework and its hide as a cover" (Vitebsky 2005: 90). Ingold (2000) makes a similar comparison between a lodge built by beavers and that of an Eskimo house, or igloo. He refers to Richard Dawkin's (1982) theory of the "extended phenotype," in which genetic effects may be situated beyond the boundary of the body of the organism. In other words, the Mongolian *ger* is an extended phenotype of the family that inhabits it, just as the beaver's lodge is an extension of the beaver itself. The herd animals also literally become an integral part of the family *ger*: The felt for the *ger* is made by family members from the wool of the sheep they herd, and the lengths of plaited rope are made from horse hair and the joins in the lattice fixed with tiny knots of horse hide. Donna Haraway (2004) writes that people are part of a dog's extended phenotype in a behavioural, ecological, and reproductive sense. Likewise, I am proposing here that the ecosocial landscape, or domestic sphere, has become part of both the herders' and herd animals' extended phenotype.

Throughout Mongolia the door of the *ger* faces to the south, or slightly southeast, to face the sunrise. In the morning, the sun streams in through the door, or the open hole in the roof (*toono*), onto specific sections of the inside walls. Interestingly, in describing placement in relation to other objects, Mongolians refer to the south as "in front" (*ömnö*) and the north as "behind" (*khoino*) because this is usually in relation to the position of a *ger*. Within any *ger*, the western side is the men's side; this side includes animal husbandry and hunting equipment. The eastern side is the women's side, containing the wife's cooking utensils and implements and access to the central stove for cooking.

[2] Divangar is a lecturer at the Mongolian State Agricultural University who assisted me for a few days when I first went into the field.

Mongolians have specific protocol for moving around within the small confines of a *ger*. One always enters towards the left, moves clockwise within, and then exits on the right side of the *ger*. The two central supports of the *ger* are not to be leant upon and nothing is to be passed between these two supports, as these poles represent the union between the husband and wife of the *ger* (another more practical reason is that the poles may become dislodged, collapsing the entire *ger*). Young animals are initially nurtured within the *ger* and nannies or ewes may be brought inside for warmth while feeding their young; however, the inside of the *ger* is primarily a human social sphere.

Within each encampment, the extended family gather in the larger *ger* (the *ikh ger*), occupied by the patriarch (or matriarch) of the family. Dogsomjav would usually sit towards the rear of the *ger* (*khoimor*), which was her position as head of the family. If Ochero was present, however, he would sit down on the floor in the position of the eldest male within the *ger* encampment, in which case Dogsomjav would move further to the women's side (the right-hand side, facing inward). Ochero's wife, Boloro, would often sit on the left-hand side of the *ger* near the door but on the bed rather than on the small stools used by guests, perhaps because she was not as at home in the *ger* of her in-laws. Lhagva would position herself on the right-hand side of the *ger*, towards the front, as the stove opened from this side and she would cook meals for her mother and any visitors. At first the intricate seating arrangements made me nervous but I began to take up my expected position naturally when visiting any *ger*. Towards the end of my stay when I was in Dogsomjav's *ger*, if her guests were not related, I felt confident enough to relinquish my "visitor seat" on a stool towards the rear of the left-hand side of the *ger* and sat on the right-hand side with the women and children of the family. I learnt my place within the social structure of the *ger*, or the human herd social structure, where gender, age, and degrees of relatedness were all parts of this structure.

THE LANDSCAPE OF THE WINTER ENCAMPMENT

In the Khangai Mountains, herders have four main campsites based on the seasons. In the following sections I describe the main winter and summer encampments of Dogsomjav's *khot ail*, as these encampments are more established and herders tend to stay at these camps for the longest periods of time during the year. In winter, encampments are spread out from one another in more sheltered locations at one end of the valley. In summer all the herders are in close proximity, in a more open, expansive location but near the river that flows through the valley, while the herds graze on the pastures nearby. The spring and autumn camps may be only a few hundred metres from the summer or winter sites, are inhabited for shorter periods of time, and tend to be more variable in location and set-up (see Chapter 6 for migration between sites).

Dogsomjav's winter camp site is near the confluence of the Khujerin River with the Selenge, the largest river in Mongolia. This camp is one of the furthest from the local administrative centre (*süm*). The location of the winter camp is crucially important, as the site is situated close to winter reserve pastures and must contain sturdy structures, which are required for housing weaker animals during winter. The grass is not overgrazed here and throughout the summer thrives, growing until it is waist high. The encampment is at the foot of a steep mountainside, which means that there are only a few hours of direct sunlight each day during winter, but the mountain provides necessary shelter from strong winds and storms. While I was there, the camp consisted of five *ger*: one for me; one housing Dogsomjav and any visitors[3]; one with Lhagva, her husband, and two children; and a fourth consisting of Ochero, his wife Boloro, and their young boy Öskö. Ochero's smaller five-sided *ger* was used as a separate storage area for food and animal husbandry equipment.

Beyond the core sphere of the family *ger*, herders constantly engage with both human and non-human beings. Directly behind the line of *ger* are shelters, mainly for young animals and their mothers. Shelters are made from black pine poles cut from the surrounding mountainsides and constructed as lean-tos with an open front. In the autumn, herding families fill gaps in the walls of the shelters with cattle dung and the floor is lined with layers of the smaller pellets of sheep and goat dung to provide an insulating layer against the cold. Such measures are crucial to protect the herd animals against the extreme conditions of winter. Supplementary hay is stored on the roof of the animal shelters. Animal dung is piled into large heaps, which is also used for lining the outside of each *ger* to retain warmth on the inside. Dogsomjav and Lhagva spend time every day clearing the herds' dung from the area of the encampment, meticulously picking up each pile, leaving the area looking as immaculate as a mown lawn. Clearing the dung provides any visitor with an immediate impression of the care and responsibility the herding women put into the domestic sphere.

One of the aspects I particularly noticed while occupying a *ger* is that, although heat and warmth do not escape, sound does carry. The shelters are

[3] I rented my own *ger* from Aagii's grandmother so that I had a certain degree of autonomy and could make use of the space for writing up field notes onto a solar-powered laptop. Davaa, my field assistant (Dogsomjav's second daughter), along with her baby, kept Dogsomjav company in her *ger*.

close enough for the herding family within their *ger* to hear the vocalisations of the herd animals and alert them to any signs of distress. During the Soviet era, herders were encouraged to move into Soviet-style concrete buildings (similar to those built in the local *süm*), but herders found they could not hear the herd animals well enough and would frequently lose their animals to rustlers and wolf attack.

The hitching post for riding horses is important as a point of connection between the encampment and beyond, as it is from here that herders arrive or depart on horseback. Hitching posts for horses are situated in front of the *ger*, so that any visitors are greeted opposite the door of the *ger*. The wind often blows from the north-west, so the wind will come first to the horses (Bayarsaikhan 2006). Upon arrival, visiting herders are expected to call "*nokhoi khorio*," or "hold the dog," to announce their arrival and to encourage a resident herder literally to restrain the dogs that fiercely protect the encampment. When people are departing, a small libation of milk may be sprinkled on the riders' stirrups, to wish them a safe journey on their travels beyond the domestic sphere of the encampment.

THE LANDSCAPE OF THE SUMMER ENCAMPMENT

The social network between herders and herds changes between seasonal encampments. The summer pastures are recognised as common ground and separate herds graze in close proximity to one another. During the winter months an encampment may migrate a long distance to another valley. Two neighbouring encampments (including their accompanying herds) may not encounter one another until the following summer, when their respective encampments are in close proximity to one another again. In the Khujerin Valley, where Dogsomjav's encampment lives, there are stands of black pine (*khar mod*). Each extended family tends to have a log cabin, built from the nearby forest (handmade with an axe and slotted together without any nails whatsoever), usually for older family members to keep cool during the heat of the midday sun. Once frosts begin, families move back into the warmth of their *ger* and move on to the autumn camp, as the wooden cabins provide no insulation or protection from icy wind conditions.

Dogsomjav's summer camp is situated on a terrace above the river and at the foot of a mountain spur, so the area is flat but elevated above both the main river valley and a side valley where the herds move off to graze for the day. From a herder's perspective it is important to have a good vantage point to observe the animals as they move out from the encampment each morning. Previously, the encampment was situated close to the river, but

Dogsomjav told me that her husband decided to move it up onto a higher terrace, as he noticed that the swampy ground near the river was damaging the legs of the sheep and goats.

The summer camp has fewer pens and no shelters for the animals but has separate hitching posts for the racehorses and drying racks for dairy products. In the summer of 2005, two *ger* were situated near Dogsomjav's log cabin. Ochero constructed the hitching post directly in front of his *ger* and another extended hitching post approximately 300 metres away, which was used exclusively for training horses for the *Naadam* season. Pens and corrals are set up with the structure of the herds in mind. The hexagonal sheep and goat pen is always near the main *ger* and one of the dogs is tied near the pen to guard against wolf attack. This sheep and goat pen is relatively small in comparison with the larger corrals built for cattle and horses, because the sheep and goats do not require as much personal space and tend to huddle together, almost on top of one another, for warmth. On the opposite side of the terrace, approximately 200 metres away, the family built the summer pen for holding the calves, and two separate corrals, one for occasionally keeping the cattle in for the night and another for catching geldings for riding. The cows are milked around the calf pen and dung is heaped into piles in a circle around the pen. These are occasionally lit, and the smoke produced aids in deterring flies while the cows are being milked. The calves' pens require some room, so that the cows can gather and have space between one another during milking.

The pens and corrals at the summer site are more temporary and are not as sturdy as the animal shelters and pens at the winter site. During the summer, the cattle and horses are routinely herded back to the encampment, as the cows or mares are brought into the domestic sphere for milking and to engage with their young, who are retained within the encampment. The sheep and goats are also herded back to the encampment in the middle of the day for milking, or to be taken down to the river to drink, then again in the evening, when they are driven into the pen as protection against wolves. The herders successfully establish each encampment as the core of each herd's home range and the herds move out to pasture and return to this core area of the encampment on a routine, daily basis (see Chapter 9, Figure 9.1).

PART II. THE SOCIAL ANIMAL

The sociality of the herd animals is an important aspect of the domestication process. An ideal herbivore for a co-domestic relationship is one that is not territorial, lives in large, wide-ranging herds, has organised hierarchies, a low response to danger, and an ability to recognise each other as individuals (Clutton-Brock 1994). All of these characteristics fit well with the social behaviour of sheep, goats, cattle, and horses.[4] This is a likely reason for their domestication in comparison to other ungulates that do not cohabit with humans.

THE SOCIAL MONGOLIAN HORSE

There are two types of horse in Mongolia, the *takhi* (*Equus przewalskii*) and the co-domestic yet free-ranging Mongolian horse (*Equus caballus*), a distinctive breed indigenous to Mongolia. The *takhi* has consistently existed in the region of Inner Asia but *takhi* populations have been inexorably pushed into marginal desert habitat from competition with the larger populations of horses coexisting with humans. In the early 1900s *takhi* were captured and taken to European zoos by train. Recently, there have been considerable efforts to reintroduce *takhi* back into two reserves in Mongolia to conserve the only representatives of "wild" horse left. Research on the behaviour of the *takhi* indicates that the social organisation is very similar to free-ranging populations, such as the free-ranging Mongolian horse, although behaviourally the *takhi* is not as placid in temperament (Boyd & Houpt 1994; King & Gurnall 2005).

Remarkably, the structure of horse society has remained relatively constant and unaltered by selection processes, whether horses are wild, feral, or domesticated. Linklater (2000) compared the social structure of twenty-one feral horse populations and captive *takhi* and found that their social organisation was remarkably similar. Social composition is divided into two types of band, based on the sex and age of individuals. One band consists of a stallion with mares and offspring (also called a "harem"). Another separate band is composed of "bachelor males," that is, an all-male group of individuals who have been ejected from their natal group. Young horses remain with their mothers for the first two or three years, whereupon males are driven away by the stallion and join other males until they establish a band of their own. The number of mares with each stallion varies in feral bands, ranging from one to twenty-six mares, whereas bachelor groups range from one to seventeen individuals. Occasionally, feral herds have been observed to form

[4] Less behavioural research has been conducted on the yak, but from personal observations yak also fit these criteria. Herders view *sarlag* as being more intelligent than Mongolian cattle, as they are very attentive to young, less placid, or what herders call "sensitive."

a number of bands into large multilevel societies, with several stallions and their respective bands tolerating one another, although within these larger associations there is still a specific social structure and dominance hierarchy (Goodwin 2002). The importance of maintaining free-ranging social bands was clearly demonstrated through the *takhi*'s difficulty in breeding in captivity. Soon after they were released into rehabilitation reserves and back onto the steppes of Mongolia, the *takhi* started to breed successfully (Boyd & Houpt 1994).

Social Structure of the Mongolian Horse

The social structure evident in feral horse populations is replicated in the Mongolian nomadic pastoral environment. Herds are separated into bands, with the stallion, mares, and young making up one band and the geldings forming a band equivalent to those composed of bachelor males. The stallion's band functions to breed young and to allow herding families to obtain mare's milk, whereas the geldings are primarily ridden. Individual geldings are ridden over the course of two or three days, and then they are released back into the herd to be replaced by a fresh mount.

The stallion is usually the leader of the band. While moving to new pasture, or travelling to a water source to drink in the middle of the day, horses move in single file. The stallion follows at the rear, as protection against wolves or other stallions. An older mare leads the daily journeys to water or in flight. Some Mongolian herders place a bell on the lead mare to help warn the rest of the herd and to distract wolves during an attack. During summer we would observe the stallion's band (*azarga*) from Dogsomjav's encampment making its way down to the river to cool in the heat of the day. The mares kept pace with one another, keeping the same spacing between each individual with the stallion always remaining at the rear.

Dogsomjav's encampment has two small herds of horses, one consisting of the stallion, mares, and young (*azarga*), and the other consisting of the castrated males (or geldings, *mor'*). In 2005 Dogsomjav's extended family had 34 horses, including 7 mares with young (see Prologue, Table P.1). Choijo's encampment had a large number of horses compared with Dogsomjav's encampment (with its two small herds or bands), and even compared with other encampments within the same river valley. Choijo's encampment had 126 horses, including 30 mares with young (see Prologue, Table P.2). Because this encampment had such a large number of horses, they were not divided up into different bands. They had five stallions with their respective mares and young, as well as all the geldings combined as one herd. This social structure replicates the social structure of a multilevel society in a wild or feral situation. Divangar confirmed this: "If horses belong to different families, then they [the herders] separate them. If one encampment has a lot of horses, then they herd this horse population together from a young age and they can pasture together." This ability to tolerate other groups is part of the non-territorial social structure of horses, which means that they can cope with differing herd sizes and changes in hierarchy depending on context.

During mare milking, tethering lines (*zel*) are laid out according to the foals from each stallion's herd. The herd is combined as one large group but subdivided into separate lines according to each of the five stallions and his associated mares and young. Foals are tied to a tethering line but the mares are free to move off and graze. Herders take advantage of the strong bond between mare and foal, as the mares stand near their foals for nursing and only occasionally wander away to graze.[5]

Close bonds can develop between individual horses and may persist throughout a horse's lifetime. Horses close to each other in rank within the dominance hierarchy groom one another and stand in close proximity to one another. Mares have preferred associates, or "friends," often siblings. In Choijo's herd in particular, one mare was always accompanied by a large piebald gelding with whom she had formed an alliance, and he was protective of both the mare and her foal. We knew to be wary of a defensive kick from the piebald gelding while approaching the mare to milk her. Many of the horses from the same herd exhibit bonding by mutually grooming one another in the evenings, or resting in a head-to-tail position to keep away flies in the heat of the midday summer sun. If horses and riders are resting, the riders may strike up a conversation on horseback, while the horses overlap necks or rest a head on the neighbouring horse's rump.

The Character of the Mongolian Stallion

Any neighbouring herder can identify a group of horses by recognition of the stallion within a herd. A stallion in Mongolia is easily distinguished because herders leave the stallion's mane to grow long, while mares' and geldings' manes are cropped short (somewhat like a uniform, where an insignia is the symbol of authority). Dogsomjav informed me that the stallion's mane is left long because it helps to drive geldings and young stallions from the mares, as other horses are afraid of the stallion's long, flowing locks. I suspect it is also because a horse with a flowing mane is aesthetically impressive to people as well.

The stallion is perceived by the herders as the "manager" of the band and is expected to take on a similar role to the herder in directing and protecting the herd. The following is an example of the Black Stallion (*Khar azarga*)

[5] See Saikhanaa and the Horses, available at www.cambridge.org/9781107000902.

"managing" his herd and fending off any geldings that may attempt to compete with him:

> The stallion is tied to the hitching post every day as preparation for the *Naadam*. At dusk I could feel the ground reverberating beneath me as I sat in my ger. I rushed outside to see the sleek, Black Stallion thundering past. He had just been released and was galloping around frantically with head extended, trying to locate the mares and young from his band. When the stallion greeted the mares he nickered and circled the band, enclosing the mares around him. Ochero said that if some of the geldings try to take any of the mares away during the day, while the stallion is tied to the hitching post, then the stallion drives the geldings away vigorously once he is released. (Field note excerpt: Dogsomjav's summer encampment, June 12, 2005)

The Black Stallion at Dogsomjav's encampment is known in the valley as a fast racehorse, and neighbouring herders traded their livestock for the stallion's offspring to train them as racehorses. Davaa informed me that her father (Dogsomjav's late husband) would ride the Black Stallion to neighbouring encampments on special social occasions, proud of the stallion's fine appearance. The stallion was good at defending the herd against wolf attack and kept as the stallion for Dogsomjav's band of mares for more than five years. The stallion was then passed on to the youngest son as inheritance. By the time I visited again in 2007 the stallion had been given to a wealthy herder, related to the youngest son through marriage, in recognition that he was a herder of 1,000 animals. Even though the stallion was past his prime, he was again a stallion (*azarga*) of this wealthy herder's mares, producing more fine racehorses.

Learning Herd Behaviour

Herders in Mongolia emphasised to me how the Mongolian horse is loyal to the herder and to its birthplace. A good, "loyal" Mongolian horse will remain at its rider's side if the rider has a fall. This behavioural trait is important for herder survival. If a lone herder falls from a horse, the herder may be too far from any neighbouring encampment to call for help or to walk to safety. I saw two instances in which Mongolian horses remained near the rider's side. When riding near a neighbouring encampment, a man had drunk so much that he had passed out and was sprawled on the grassland steppe; his horse patiently waited nearby. On another occasion, my field assistant Baigalaa was cantering beside me when the horse she was riding tripped and she somersaulted over the horse's head. She was shocked but unscathed, and she remarked soon afterwards that the horse had not run away but waited until

Table 3.1. Timeline of Mongolian horses

Structure or influence	**First year (*unaga*)**	**Second year (*daaga*)**	**Third year (*shudlin*)**	**Fourth year + (*khyazaalin*)**
Social structure	Males and females: remain with mother in *azarga*		Males: if castrated, in herd with other geldings	Stallion replaced every four years
			Females: remain in natal herd with mother	Females: pregnant with young
Human influence	Handled and tethered during milking	Males: castrated in autumn, or kept as future stallion	Used for herding and travelling to other locations	Milk obtained for *airag*
		Males and females: broken in for riding purposes		
	Not eaten in first year		If unproductive, killed for food	

she had recovered and got back on again. The horse is driven to stay with its human companion, rather than fleeing. It was only after these instances that I understood what herders meant when they described the Mongolian horse as being "loyal."

Mongolian herders allow horses to manage themselves while grazing throughout the day. They are free to roam wherever they choose, and much of a herder's time with regard to the horses is spent in finding the location of the herd each day. Horses are rounded up only during the milking season so that herders can catch foals and milk mares or, in the case of the geldings, catch an individual for riding. In these herding instances, the herder on horseback takes the place of the stallion in that the person rides behind the herd (I saw no instances in which the herder took the position held by the lead mare at the front). Mongolian herders restrict the social behaviour of the horses within a herd to a minimal extent. They take advantage of the horse's normal behavioural repertoire to carry out activities, such as milking or riding (see Table 3.1 for stages in Mongolian horses' lives).

THE SOCIAL MONGOLIAN COW

Mongolian cattle, *sarlag*, and hybrids (*khainag*) are herded together in the mountainous regions of Mongolia. The behaviour of Mongolian cattle and *sarlag* (the Mongolian word for yak) are similar, although Mongolians recognise that the indigenous breed of Mongolian cattle is more placid and has less of a tendency for being "wild" (*zerleg*). Mongolian cattle are docile and graze together in small subgroups, whereas *sarlag* are more reactive and slower to habituate to humans or new circumstances (Tumerjav 2003). *Sarlag* seem to vary widely in their temperament and behaviour; for instance, an individual may be cranky (*aashtai*) or smart (*sergelin*). Women milking *sarlag* describe them as "kind," because they are "good mothers" and willingly adopt young that are not their own (see Chapter 6 for the topic of surrogate mothers). According to Tumerjav (2003), their mothering instinct is well developed and they may fiercely defend members of their own herd. Choijo informed me that *sarlag* are good at defending themselves against wolf attack and are less frequently killed by wolves than other herd animals. I saw members of the herd actively chase away any dogs that came near. Herders and herd animals mutually adjust to the various characters and temperaments of individuals within a herd.

Social Structure of Mongolian Cattle

Adult cattle are free ranging in Mongolia and choose their own pasture within a home range surrounding each seasonal encampment. Most of the human contact with cattle is with females during milking, or when oxen are caught to pull wooden carts for tasks, such as collecting firewood or moving possessions to a different encampment. During the summer milking season, adult cattle are divided into two main herds (similar to the horse herd), one of which consists of the dairy cows and bull and the other of which consists of those cows (such as *gunj*) that have no young and all castrated male animals. This means that the dairy cows return twice a day for milking, while the rest of the herd remain grazing. Young that still nurse remain separate from the rest of the cattle; this group consists of first-year calves and some second-year *byaruu* that still nurse from their mothers. They are contained within a pen during milking and at night, but they are free to graze near the encampment, once the adult herd is out of sight, during the middle of the day. If they are not kept separate from the rest of the herd they will run to their mothers and nurse, leaving no milk for the herding family.

During milking times, I noticed that the older, more dominant cows routinely stood closest to the gate of the calves' pen, while the young heifers (*tulgar gunj*) stood towards the outer edges of the cow herd. If the dominant cows were released from their hobbles earlier than the more subordinate cows, they would bump into or butt the subordinate cows as they walked past. During milking, to avoid any conflict between the cows, we would

Table 3.2. Timeline of Mongolian cattle

Structure or influence	First year (*togal*)	Second year (*byaruu*)	Third year + (*gunj/gunjin*)
Social structure	Males and females: socialise with other first-year calves	If not nursing, graze with rest of herd	Males: graze with herd
	Bonds with mother during nursing	Some still nurse from mother, kept with other *byaruu*	Females: pregnant or already with own young
			Graze with other cows
Human influence	Often assisted to suckle from mother soon after birth	Males: castrated in autumn or kept as future bull	Castrated males: trained to pull loads with ox cart
	Handled during milking times		Female with young: hobbled and milked daily
	Not eaten in first year		If unproductive, killed for food

either guide the cow past other hobbled cows still waiting to be milked, or leave the dominant cows to wait until the more subordinate cows that were on the periphery of the herd had finished nursing their young. The cows rarely threaten humans with their horns (if they have any) and generally avert their head away in a submissive gesture if anyone passes close to their head, an indication that humans are treated as dominant individuals within the cow's social hierarchy (see Table 3.2 for stages in the lives of Mongolian cattle).

The Character of the Mongolian Bull

Mongolian bulls (including both *sarlag* and Mongolian cattle) seemed placid in comparison with European breeds of bull that I had previously encountered. I previously had the perception that bulls can become uncontrollable and could charge indiscriminately at humans. This difference in the behaviour of Mongolian bulls could be attributed to a variety of factors: Mongolian bulls may be innately placid; they may become passive as a result of the way Mongolian herders interact with them; or perceptions of the innately aggressive tendencies of bulls may be unfounded. I suspect it is a combination of all of these factors. I observed that Mongolian bulls do not become agitated through social isolation and roam freely to mate with receptive cows.

The dominant bull (and the big, heavy oxen) behaves submissively towards herders, even young children (also see Lott & Hart 1979). Bulls, like all cattle in Mongolia, are handled by humans from the time they are born. When young, they are released from the calf pen to nurse from their mothers and maneuvered back into the pen afterwards, during both the morning and the evening, throughout much of the year. This means that during their development into adults, a young bull learns to understand the vocalisations and body language of the herders and habituate to human presence. Even young children, like Öskö, could move a bull with a threat gesture by brandishing a stick, or idly kick the bull on the rump while the bull lay amongst the cows during milking.

The bulls do, however, become highly competitive and antagonistic towards one another when cows come into oestrus, during the breeding season.

> Dogsomjav's bull (Khurnee) has become a problem during milking because he no longer sits placidly amongst the cows. He paws the ground and bellows right beside us while we are engaged in the morning's milking. When the bull is agitated, pawing the ground and bellowing, he still does not become aggressive towards humans. I stood directly in front of him and he stopped his impressive threat display immediately.
>
> After a couple of weeks of build up, the bull from this encampment and the neighbouring bull started physically fighting and butting. Previously, both bulls performed visual displays to one another (at a distance of more than 100 metres away), not only pawing at the ground and flicking up clods of soil but also rubbing their heads and necks in the dirt. Today Dogsomjav's bull won a physical jousting bout against the neighbouring bull but he has bloody gashes on his thick neck. The neighbouring Mongolian bull employs a high-pitched bellow, a bit like a zebra, as he walks past with his herd. A couple of the young men from the neighbouring encampment tried to chase Khurnee over to joust with their bull, while we were milking, but Dogsomjav yelled for them not to chase him. Like breaking in horses, it seems to be an entertaining spectacle for young men here to watch the bulls fight one another. (Field note excerpt: Dogsomjav's summer encampment, June 4, 2005)

According to an old herder I spoke with, Mongolian bulls tend to compete with other bulls in early spring and summer when they are strong from good pasture. *Sarlag* are stronger at fighting later in the season because they are more resilient to colder temperatures. The agonistic encounters of *sarlag* are similar to Mongolian cattle, but *sarlag* make a series of panting grunts in quick succession and exhibit visually intimidating tail-waving with their bushy tails.

Bulls (*bukh*) and Mongolian Wrestling (*bökh*)

Mongolian herders observe the behaviour of cattle closely and have integrated the behaviour of the breeding males' agonistic interactions into a ritualised sporting activity: Mongolian wrestling (also see Chapter 7). Tomikawa (2006) also refers to the wrestler's posture as being in imitation of a bull. He adds that it represents breeding stock in general, including stallions and camels. According to Tomikawa, one reason for this symbolic reverence of breeding stock is that they have the ability to protect the rest of the herd from predators, such as wolves. As I explain in Chapter 4, the breeding males contain the spirit or essence of the herd as a whole and are also representative of the strength and status of the herding family.

Male cattle compete with one another for status in the dominance hierarchy by pawing the ground, butting, and mounting. Their display is highly ritualised and designed as a visual and auditory performance towards the opposing male. Each bull paws the ground, flicking clods of dirt into the air, and makes a deep, resonant bellow; this then escalates to rubbing the head and neck, or digging horns into the dirt. When physical contact occurs, the two opponents lock heads, pushing at one another, which may last to the point of exhaustion. This is an effective means of assessing an opponent's weight: an "honest" means of judging relative fitness. Cattle may remain locked head to head for long periods of time, with the occasional quick burst of activity in which one twists and lacerates the other with a horn, or backs off and assumes defeat.

This is remarkably similar to the technique used in Mongolian wrestling, which is ritualised, involves physically strong, young males, and is a competition for social status. The wrestling bout is preceded by a preliminary eagle dance (*devekh*) and loud thigh-slapping, allowing an opponent to assess the other's fitness even before they have begun to wrestle. The head-lock stance, in which opponents remain shoulder to shoulder, is very similar to cattle remaining locked together head to head. As there is no time limit, wrestlers lock arms for long periods of time, assessing the strength of the other with only a burst of action at the end when one is hoisted up and then pushed to the ground. There are no weight categories in Mongolian wrestling, so the biggest wrestler usually wins (as is the case with cattle). As the comparison between Mongolian wrestling and fighting between competing bulls demonstrates, herders observe the behaviour of the herd animals and incorporate such actions into their own behaviour.

THE SOCIAL MONGOLIAN SHEEP AND GOAT

According to Hemmer (1990), the intensity of movement and liveliness of domestic sheep and goats has become reduced. These behavioural changes are particularly evident in sheep and goats living in intensively farmed situations, resulting in exaggerated changes to morphology, physiology, and behaviour. Some free-ranging, isolated sheep populations, however, such as the Soay from St. Kilda, have retained "wild-type" behavioural characteristics (Clutton-Brock & Pemberton 2004; Jewell, Milner, & Boyd 1974).

Goats and sheep in Mongolia are still highly mobile while moving and grazing on mountain slopes, as well as keenly alert towards the potential threat of predation by wolves. Sheep and goat herds in Mongolia are more closely directed and monitored by herders than the cattle and horse herds, but they still graze on a variety of vegetation, moving to different pastures frequently, and actively engage in seeking out new vegetation to forage upon for themselves. I suspect that, like the free-ranging sheep on St. Kilda, the fat-tailed sheep in Mongolia have retained wild-type characteristics[6] from factors such as external selection pressures existing in the form of extreme weather conditions, a need to search for appropriate vegetation, potential predation from carnivores, and variable terrain. Unlike many breeds of sheep and goat, the endemic breeds of fat-tailed sheep and cashmere goats in Mongolia could be referred to as "free-ranging" breeds, as opposed to the "intensively farmed" breeds. The free-ranging Mongolian breeds have retained many behavioural capabilities necessary for survival in a harsh environment through being able to occupy an area encompassing a large home range, and they interact both cross-generationally and cross-specifically (much like the Mongolian herders do themselves).

Multilevel, Cross-Species Herds

Tani (1996) argues that, in contrast to a so-called wild flock in which there is a vertical mother–offspring relationship, herders organise the group so that it is horizontal and co-generational, with herds divided into different age cohorts. This system was particularly adhered to during the Soviet era within the herding *negdel* in Mongolia. Dogsomjav explained to me that, as a child, her family nurtured the cohort of lambs and kids that were born that year for the entire community, while adult sheep and goats were herded by other herding families. This structure has since changed; it is only for the first few weeks, until the lambs and kids are strong enough to follow the herd

[6] This reference to "wild-type" sheep, however, perpetuates a dichotomous view of domestic versus wild.

out to pasture, that newborns are separated from the rest of the herd. Now that herders can choose the structure of the herds for themselves, for most of a sheep or goat's life the herds are herded as multigenerational, multispecies flocks.

The basic social unit of sheep and goats is the female with her most recent offspring, followed by a more loosely associated unit of related sheep, such as the addition of a yearling. Herders make every effort to maintain the basic social unit of sheep and goats by assisting with the bond between mother and young. When lambs and kids are just born, the herders allow the mother time to lick her newborn clean, as this aids in the recognition of her young by smell. The herder will then remove the newborn to shelter with the mother in order to provide protection from wolf attack, or adverse weather conditions. The following day the newborn is placed with all other kids or lambs that have been born that season but is matched with its mother during feeding times. From a very young age the two species – sheep and goat – interact, butting heads against one another and playing together. Mongolian sheep and goats have a highly enriched environment in which to learn, with the ability to roam over large areas and interact not only with individuals from the same age group but even cross-specifically. All learn from their continual engagement with one another: sheep, goat, and human.

The Herder of Males

One herder, within a group of herding encampments, is assigned with herding all the breeding male sheep and goats from neighbouring encampments for most of the year, essentially forming a bachelor herd of only breeding males. This is similar to populations of ibex or argali, which form small splinter flocks of males. Herders restrict the breeding ram and billy to entering the herd for only a short period of time, approximately two weeks, to ensure that all lambs and kids are born over the same period in spring.

> Over the previous three years, Dogsomjav's encampment had a billy goat (*ukhna*) with no horns, resulting in most of the herd having no horns. They recently castrated the *ukhna*, as the extended family replace their breeding males every three years to avoid inbreeding within the herd. Dogsomjav pointed out a particularly small and sickly kid, that the family had affectionately named *Missmaa* (meaning "Little Miss"), as an example, and told me that some of the younger goats are having kids that are too small. The family recently acquired a two-year-old billy (*borlon ukhna*) from a neighbouring administrative district with impressive horns that grow together, vertically, and a blonde, sprouting forelock that reminds me of a punk hairstyle. Although still small, he has started mating some of the females. Because it is the wrong

> time of year for mating to begin, Ochero separated him from the herd and took him to a herder that tends only breeding males. This herder usually only takes animals from herd encampments within the valley at the start of spring, [because if he takes them] any later the strongest animals may fight and compete with one another to the point of exhaustion. Yet he took this *ukhna* late in the season as he is still small, so is not large enough to be much competition for other males.
>
> When we arrived on horseback at the breeding-male herder's pen, the old herder said to Ochero not to bring the horses too close because sweat[7] from the horses may arouse the breeding males and they could start butting one another. The old herder emphasised to me that the breeding males are "particularly sensitive." Ochero and the old herder took the young *ukhna* by the horns and tied him to the pen, so that he could become accustomed to his new environment. On initial encounters males may fight to the death, so they are initially tied to prevent them from harming one another. When the males persistently butt one another, the old herder treats them with *aarts* (curds from sour milk) and reads to the animals from sacred Buddhist text. He added that it is only in autumn that the males become more difficult to manage because this is the prime mating season. (Field note excerpt: Dogsomjav's summer camp, June 5, 2005)

In exchange for herding the breeding males throughout most of the year, the old herder receives a kid or lamb in return. Dogsomjav's young billy was later attacked and eaten by wolves, so Dogsomjav borrowed another billy from a neighbouring encampment to breed with the female goats during a ten-day period in autumn. At times of the year when the billy and ram are part of the flock (the only males that are not castrated), the billy and ram will graze towards the front but they are unlikely to take the role of lead animal as their primary focus is on mating with females.

Social Structure of the Mongolian Sheep and Goat Herd

When the flock is grazing, the herder takes a position at the front of the flock to guide them, taking the lead role, or sits in the middle of the herd when the flock is comfortably grazing and spread out. Wolves use the strategy of moving in from behind, as the weaker individuals will tend to be towards the rear of the flock. To protect the flock, herders tend to move the herd along from behind when moving from one area of pasture to another. When the herd is moving towards fresh pasture, the goats are more nimble on their feet and tend to lead the herd with the sheep bunched together towards

[7] See Chapter 7 for more details on the significance of the sweat from horses during race training.

the rear. Both individual goats and sheep retain the same relative position within the flock while moving and grazing. Even within the pens or shelters, individuals within the flock will retain the same positions night after night.

Boloro explained to me the benefits of herding sheep and goats together while she and I were out herding one day. Sheep are content to stay on the more gradual slopes at the foot of a mountainside and graze on whatever is available nearby. Mongolian sheep are more stable than goats; once they arrive at new pasture, they quickly settle and begin to rest. Mongolian goats are clever, lively, and responsive to external stimuli. Compared with Mongolian sheep they are better able to fend for themselves and are particularly sensitive to sounds and smells, making them highly adaptable when searching for new pasture. Goats are more exploratory and inclined to move quickly from one patch of ground to the next. They are opportunistic grazers and are adept at finding good pasture. Mongolian herders vocalise to redirect the goats to prevent the lead goats from moving too far away and scattering into smaller groups.

With the first shoots of new growth in spring, goats are difficult to retain within a flock, as they are highly motivated to seek out soft flowers and shoots on steeper slopes. They are adept at standing on their hind legs to graze on small trees and bushes. In spring and early summer the black pines have new needles that are edible, and I observed goats eating the soft needles by standing on their hind legs. The sheep tend to graze nearby, although I did witness a few sheep also standing on their hind legs to get at the needles, possibly learnt from the foraging behaviour of the goats. Because sheep and goats can learn from one another to better survive the harsh environmental conditions, it is beneficial to structure the herd in a cross-species, cross-generational herding structure (see Table 3.3 for stages in the lives of Mongolian sheep or goats).

Mongolian herders have an expert knowledge of herd behaviour, especially the sheep and goat herd, because the herders spend such large amounts of time sitting, observing, and communicating with animals. An example of how herders use this knowledge to their advantage is in their subtle alteration of the herd's rate of movement. During the morning and evening, sheep and goats chew their cud while they are resting, but if they are not chewing it is because they are lacking energy or are actually sleeping. During midday and towards evening the flock tends to become better established and more cohesive, and the flock may rest without chewing their cud. When a herder wants to rest for a couple of hours in the middle of the day, she may keep them moving throughout the morning, so that the sheep and goat herd needs to rest and ruminate by midday.

Table 3.3. Timeline of Mongolian sheep and goats

Social structure or influence	First year (*khorag/ishig*)	Second year + (*tylög/borlon*)
Social structure	Males and females: socialise with other first-year lambs and kids Bonds with mother during nursing and when old enough will graze with herd	Main herd graze together; breeding male kept in separate herd Females: may become pregnant and give birth to young
Human influence	Herders assist young to suckle from mother soon after birth Females: Handled during milking times Males: castrated in spring or kept as future ram or billy.	Females: milked for one month in summer Goats: combed to obtain cashmere Sheep: clipped to obtain wool for making into felt
	Not eaten in first year	If unproductive, killed for food

Humans apply human-like characteristics to other animals, a concept known as anthropomorphism. Not surprisingly, given that humans are inclined to apply human attributes to other beings, when animals engage with humans, the animals also often react as if they are encountering other members of their own species (unless they interpret the human as a potential predator). This behavioural phenomenon in non-human animals is referred to by Walther (1991) as "zoomorphism," with anthropomorphism as its human form. According to Walther, one of the reasons why humans and ungulates are able to communicate with one another is because each will anthropomorphise or zoomorphise the other. When ungulates of a non-territorial, gregarious species, such as horses and cattle, zoomorphise humans, the human is then able to take on the role of another member of the herd.

Vinciane Despret (2004) describes emotional relations in the practice of domestication as "anthropo-zoo-genetic," a practice that constructs both human and non-human. She provides an apt example through the ethologist Konrad Lorenz and his relationship with his geese. Lorenz and his geese, "in a relation of taming, in a relation that changes both identities, have domesticated one another" (Despret 2004: 130). According to Despret, Lorenz's experience of bonding with a goose is not only an anthropomorphic or zoomorphic experience, but "the whole experience is a shared experience, an experience of being 'with'" (Despret 2004: 130, 131). Cross-species encounters are formative; recognising the "self" in the "other" is not merely projection, but rather has the potential to be a synergistic encounter in which both parties learn from one another.

Herders engage with the herd animals' sociality in a "zoo-anthropogenetic" manner. Ingold draws attention to the fact that in some cultures animals do not just hold metaphorical or metonymic meaning in people's lives but "they participate as real-world creatures, endowed with powers of feeling and autonomous action, whose characteristic behaviours, temperaments and sensibilities one gets to know in the very course of one's everyday practical dealings with them" (Ingold 2000: 52). Instead of drastically altering the herd animals' surrounding environment, or the basic structure of the herd, Mongolian herders take advantage of the social structures that are already an inherent part of the animal's sociality, still evident in wild or feral representatives of the same genus. The sociality of the herd animals is crucially important to cohabitation with humans, and incorporated within this is the social "landscape" of the herding encampment.

CONCLUSION

This chapter has examined how it is not just Mongolian herders who make up herding society; herd animals and the landscape are actors in, and an active part of, this sociality. Ungulates are social animals. Each herd is structured according to a hierarchy with members that are dominant and subordinate. The herds in Mongolia are essentially structured in a similar manner to the social structure of herd animals in other parts of the world in wild or feral populations. The structure is not necessarily linear, like a pecking order; it is complex, and it changes depending upon the animal's age, breeding status, and even context. Mongolian herders align themselves as part of this social structure, by entering into the hierarchy as a dominant member of the herd, taking the lead role. Members of the extended herding family are readily accepted as an integral part of the herd animal's sociality.

The landscape facilitates this sociality, functioning as a core area within which herders and herd animals interact. In zoological terms, this would be the core of an animal's "home range." Perhaps the simpler use of the word "home" is a word that any social person can identify with. Every day the herders engage with others within the surrounds of the herding encampment, not just humans from their extended family, or visitors like me, but also with individual animals from each herd, whom they interact with daily, throughout their lives. The *ger* functions as the core of the domestic sphere, where young humans and herd animals are nurtured. Dogs guard the encampment, providing protection against strangers stealing any of the valued livestock, or against wolves that dare to invade this sphere to attack the sheep and goats. Female animals are particularly drawn to return to this core area, driven to nurse and bond with their own young. Herders mainly leave the domestic sphere of the encampment accompanied by the herd animals, on the back of a horse to guide and protect the sheep and goat herd while out grazing on the grassland. It is necessary to take into account the ecosocial landscape herders inhabit and how this affects the engagement between Mongolian herders and the herd animals who interact with them as part of the herder–herd animal co-domestic relationship.

4

NAMES, SYMBOLS, COLOURS, AND BREEDING

INTRODUCTION

Mongolian herders' description of herd animals is complex and encompasses a large number of terms based on an animal's physical appearance, behaviour, sex, age, and the characteristics of a particular individual. This chapter provides a description of the selection of specific animals for breeding and how the categorisation of animals is a key factor in the ability of human and other animals to coexist with one another. Mongolians class animals at the species level, which is "a combination of appearance and temperament, and derived from close study of their habits and mental and physical abilities" (Humphrey with Onon 1996: 98). Animals are also classified at the level of the herd but, within a herd, individuals may be marked to indicate the family to which they belong. Lastly, animals are identified at the level of the individual.

Herders assess the animals as having "good" or "bad" characteristics, just as humans have specific behavioural or physical characteristics. Some are given generic descriptive terms that can be used for any individual of that species with the same type of colour or pattern. Others are given unique names that describe the characteristics of the animal's personality with a name specific to that individual. This identification of particular individuals is extended to the selective breeding process. The herder ultimately chooses whether an animal is retained for producing young; castrated and utilised for riding or pulling carts; or culled because it is unlikely to survive or produce viable offspring. Within this chapter I outline the herders' influence on the selection of an animal followed by the nomenclature herders use when identifying herd animals.

PART I. BREEDING

In Mongolia the line between what is "natural" selection and what is human-influenced selection is not clear. There is no point at which the herd animals are "made" by the herders; instead, both humans and non-humans have an influence upon each other, forming extended phenotypes of one another. Unlike ungulates in intensively farmed situations, Mongolian herd animals have the ability to survive harsh environmental conditions with less reliance on humans for food and shelter. Co-domestic animals in Mongolia forage on vegetation similar to the plants foraged on by their wild progenitors, and they roam freely when foraging. The herds are still subject to predation by wolves and have to be hardy enough to survive the extreme cold of the Mongolian winter. Nonetheless, herders do exert crucial influences upon the herd. Mongolian herders select males for breeding or castration. These males are selected not only on the basis of physical appearance but also upon their behaviour within the herd. The selection techniques used by Mongolian herders have developed over centuries of experience, relying on what Mongolians refer to as the "folk" or *arat* method of breeding livestock.

MONGOLIAN BREEDS

Each province within Mongolia has breeds that are well adapted to the type of environment within each area, whether that consists of desert–steppe, grassland–steppe, or mountain–steppe conditions. The different breeds are generally named after their region of origin, or their predominant coat colour. Sheep, followed by goats, have a large variety of breeds spread across Mongolia's different provinces. In all five kinds of ungulate, the main breed is the well-adapted *Mongol*, with only a small proportion of the stock consisting of other breeds (see Bynie 2002). Indigenous breeds have adapted to the extreme climatic conditions of Mongolia so that they are capable of safely absorbing high levels of sunlight and are protected against extreme temperature fluctuations. Mongolian livestock develop a thickened skin and layers of undercoat to survive through the winter. With thicker winter hair and a thick layer of fat (obvious in the fat-tailed breeds of sheep), Mongolian breeds are capable of surviving in particularly frigid conditions.

Endemic Mongolian breeds of sheep, goats, cattle (including yak), and horses are all remarkably variable in coat colour. As Epstein (1969: 101) attests in reference to the Mongolian horse, "coat colour is very variable, [with] practically all colours and colour patterns found in domestic horses being encountered in the Mongolian [breed]." This variability in external features was particularly noticeable to me in Mongolian sheep, as my previous experience was with intensively bred sheep, such as Merino or Romney, which are mainly a homogenous white colour. Mongolian indigenous sheep have remarkable facial patterns, such as eyespots, blazes, or flecks of white on their ears. This large degree of colour variation in native Mongolian stock may have been caused by many factors, both environmental and socio-cultural.[1]

The Father of the Herd

In Mongolia, herders make many references in relation to "father" or "grandfather," for example, referring to the God of the land as Grandfather (*Ovgon*). The breeding male animals (or the "father" animals) are the spirit or essence of the herd, which continues on from one generation to another. The herders' main influence on the selection process of a herd is through the selection of male animals, as there is usually only one male animal that is not castrated within each herd. The herder selects the breeding male based on assessment of physical and behavioural characteristics, as well as the individual animal's potential to produce offspring. Mongolian herders are proud of the breeding male animals and will point them out to visitors to admire. To Mongolians, it is the quality of the breeding males that ultimately determines the quality and strength of the herd as a whole. The stallion, for instance, is crucial for maintaining the physical and spiritual qualities of the herd and is referred to as the "father" of the herd. This is because the stallion is literally the father of all the young, as all other males are castrated or herded separately (refer to Chapter 3). The stallion is used as an example to young male herders, because of the stallion's skills as a "manager" and protector of the mares and young.

Many pastoralist societies are recognised as retaining remarkable memories and strong oral traditions for recounting the genealogy of their herd animals (for examples, see Fukui 1979, 1996; Osmaston et al. 1994). Mongolian herding families do not make written records of the pedigrees of the herd animals, but they are able to mentally recall the genealogy of their horses and cattle for a number of generations. As an example, Ganbar described the familial relationship of the *Khaltar azarga* (Dirty Stallion; see Figure 4.1). He recalls two prior generations and the stallion's siblings, and then the stallion's own band of mares and respective young.

Many of the family encampments within one valley are interrelated through marriage. Likewise, many of the cattle within one valley are interrelated. Although the coat colouration and pattern on the animals is highly

[1] For interesting research undertaken elsewhere on factors influencing coat colour variation in African cattle, see Finch and Western (1977) and Fukui (1996).

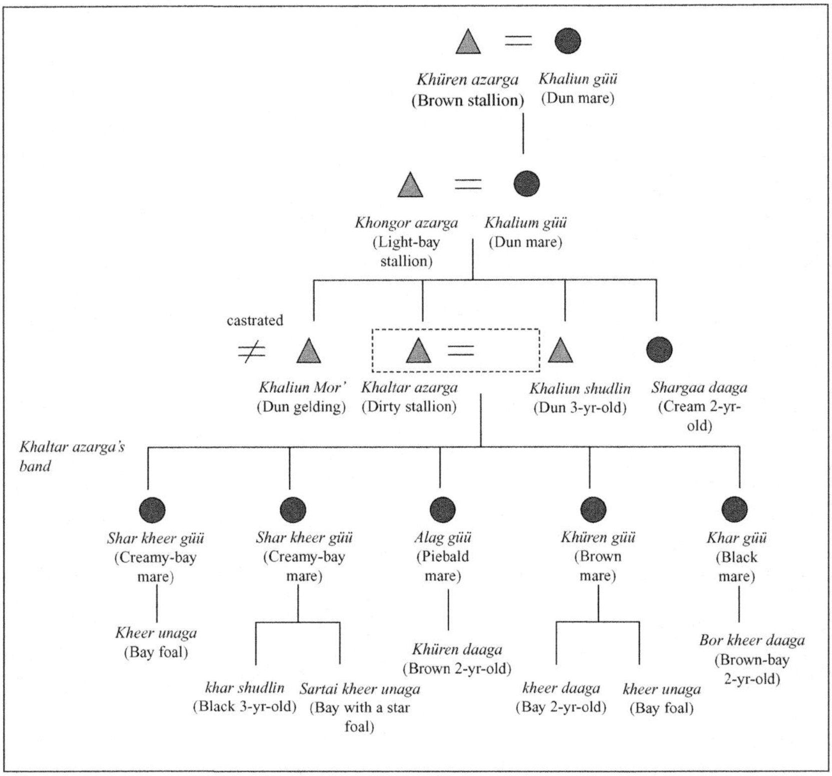

Figure 4.1. Genealogy of a Mongolian horse: *Khaltar azarga,* described by Ganbar, Choijo's encampment.

variable, these colours and patterns reoccur throughout the herds within the same valley. On occasions when I observed the milking at neighbouring encampments, many of the individual cows or calves had similar colours or markings to individuals within Choijo's herd. I noticed that Sumya's black-and-white bull (*tarlan bukh*), with distinctive facial markings, would wander to one of the neighbouring herds. During milking time at this neighbouring encampment I noticed that many of the calves had similar facial markings to Sumya's bull and the women confirmed that the *tarlan bukh* would "visit" their cows during milking.

When I asked Choijo about the names of the cows within the encampment, he quickly replied that "Naraa knows the cows." Women tend to know the different generations of cows, whereas the men know the generations of horses; this is due to the respective working roles of men and women. The herding women are sometimes unsure of the sire of particular cattle, as neighbouring bulls may mate cows from their herd. This means that, in the

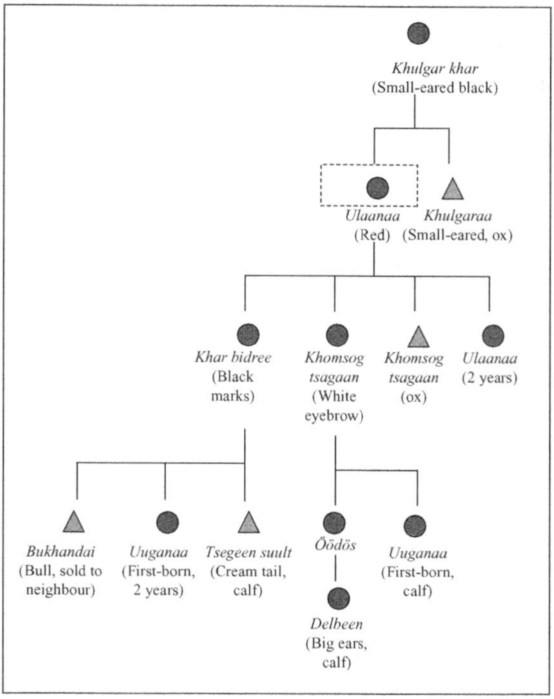

Figure 4.2. Genealogy of a Mongolian cow: *Ulaanaa,* described by Lhagva, Dogsomjav's encampment.

case of cattle, the relationships are traced through the female line, as it is the cows that the women interact with on a daily basis. As an example, Lhagva related the familial relationship of the cow *Ulaanaa* (Red; see Figure 4.2).

The main intervention on the part of the herder is through the selection of one breeding male from the herd and the castration of all other males. Although castrated males are not used for breeding purposes, they play crucial roles within a herding encampment, because geldings and oxen grow to be large in stature, yet tame, so they make good animals for riding or traction.

CASTRATION

A particularly important occasion during the herding year is the annual castration of male herd animals. Castration (*khungulakh*) is a key part of the selection process, as it is the main means by which herders select those males that will produce further offspring and those males that will be used for other purposes (see Table 4.1 for Mongolian terms regarding a herd animal's breeding status). Castration is also a means of control, particularly for riding

Table 4.1. Terms for male and female herd animals, based on breeding status

Breeding status	Horse	Cattle (incl. Sarlag)	Sheep	Goat
Breeding female	*Güü*	*Ünee*	*(Em) Khon'*	*(Em) Yamaa*
Breeding male	*Azraga*	*Bukh*	*Khuts*	*Ukhna*
Castrated male	*Mor'*	*Shar*	*Ireg*	*Serkh*

or traction animals such as geldings (castrated male horses) and oxen (castrated male cattle). Lambs and kids are castrated in the spring before they become too large and begin feeding on grass. The cattle and horses are castrated in the autumn of their second year (as *byaruu* or *daaga*, respectively). Similar to the slaughter of an animal, there are specific rules to abide by so that the animals do not suffer unnecessarily from the castration process (also see Chapter 10). The following was an account of the "no blood" method of castrating lambs and kids:

> Today is the third day of the month on the Mongolian calendar, an auspicious day for the castration of this year's lambs and kids. An offering bowl, piled with dried dairy products, fried dough and sweets, and burning incense, is laid out near the castration site. The act of castration is performed by two men, one holding the animal still, while the other performs the operation. Women and children select out the female and male lambs and kids. They hand the male animals to one of the men, who holds the lamb or kid still on its back with limbs outstretched in the same manner as adult sheep and goats are held for slaughter.
>
> Castration is performed with a small scalpel. The herder finds the testes, makes a small incision [in the skin] and pulls the entire testicle out. This is then repeated with the other testicle. He then pinches the wound from the incision closed and places the testicles in a bucket of milk mixed with water. The kids make a painful bleating sound then stagger to their feet upon release. The weaker animals lie down on the grass recovering. After all animals have been castrated, the family eats some food from the offering bowl. The testicles are taken inside the *ger* to be cooked immediately. Herders believe the testicles must be boiled and eaten straight away, so that the lambs and kids will recover quickly. Everyone involved eats all the testicles (which reminded me of the taste of egg but have a similar texture to oysters). Usually dogs are given anything that is left over but castration requires special respectful practices. The leftover broth is thrown in a location where the dogs cannot find and eat it. In this instance Lhagva poured the contents on the roof of an animal shelter. Milk products are not to be given away for seven days after castration. Davaa informed me that a relative took milk products back to the city only a

few days after castration and two of their horses died soon afterwards. (Field note excerpt: Dogsomjav's spring encampment, May 11, 2005)

It is evident from this excerpt that herders are aware of the potential impact the castration process may have on the animals, and measures are taken to prevent any negative consequences from the surgery. If the testicles are mixed with "white food," such as cow's milk, the herd is thought to have greater reproductive success. As the herders can do little to assist the animals themselves in their recovery, the family makes every effort to conduct the castration at a time of year when the animals are most likely to survive; the herders believe that the outcome is dependent on respectful behavioural practices before and after the castration event. To Mongolian herders, the reproduction of herd animals is crucially important. They recognise that to deny these male animals the ability to reproduce is significant, so the correct rituals and procedures must be abided by to ensure that the herd as a whole will continue to reproduce and prosper.[2]

CROSS-BRED COWS

Mongolian herders have an extensive knowledge of the ancestry of their herd. The complexity of the breeding process between *sarlag* and *Mongol* cattle is a good example of the herders' in-depth knowledge of herd genetic composition. In Mongolia one-fourth of the cattle are yak, or *sarlag*. Dogsomjav's herd consists entirely of Mongolian cattle, as the altitude is too low for *sarlag* to reproduce well. Choijo's encampment manages a mixed herd of *sarlag*, Mongolian cows, Mongolian cow–yak crosses (*khainag*), and second-generation (*ortoom*) or even third-generation crosses (*usan ortoom*).

Herders in the Khangai prefer *sarlag* without horns, as they are easier to handle, and there is a general view amongst herders that these animals lose a large amount of body heat in winter through their horns. In the Khangai Mountains, *sarlag* tend to have no horns (90 percent of *sarlag* in Arkhangai are polled; see Magash 2003). Dogsomjav's *Mongol* cattle did sometimes have horns, but many of the horns were reduced to stumps because they had fallen off in the extreme cold. It is only hardy Mongolian breeds of cattle that can survive in such conditions throughout the year, although *sarlag* and hybrids tend to fare better.

Most regions that breed yak–cattle hybrids generally mate yak cows with bulls from cattle (Epstein 1969; Magash 2003), yet in the Arkhangai region the herders tend to have *sarlag* bulls mating with a small number of

[2] See Medicinal Treatment, available at www.cambridge.org/9781107000902.

Mongolian cows. When a *sarlag* bull mates with a *Mongol* cow, the offspring that is produced is called a "*naran khainag*" (or "sun hybrid"). The less frequent cross of a *Mongol* bull with a *sarlag* cow produces a "*saran khainag*" (or "moon hybrid"). If the hybrids are male, then they are sterile and are trained as oxen. Hybrid oxen are used for carrying loads in preference to *sarlag* oxen because they tend to be larger, stronger, and what herders describe as "tame" (*nomkhon*). Mongolian herders prefer female *khainag* because they produce more milk than purebred *sarlag* and have thicker and creamier milk than *Mongol* cows.

Hybridising *sarlag* with *Mongol* cows requires a good knowledge of pedigree, because after two generations of hybridisation the crosses produce offspring that Mongolians call *ortoom*. Herders do not favour *ortoom* after the second generation of hybridisation because the females are unable to reproduce, resulting in both male and female *ortoom* being eaten as two-year olds (during the autumn cull). If an *ortoom* cow again produces offspring, this generation is then described as "*usan ortoom*," which means that the lineage is "like water" (refer to Figure 4.3).

SELECTION

The "folk" (*arat*) means of selection is based on the condition (*shinj*) of the animal, its ability to produce young or milk, the quality of its offspring, and its ability to handle harsh conditions. Every three to four years, herders change the breeding males within their herds to avoid inbreeding and to maintain the quality of the herd. The details given here provide an example of the features herders use when choosing a horse for breeding purposes. The criteria vary, depending upon the kind of herd animal.

The selection of the stallion is crucially important, as this individual represents the quality and status of not only the herd but the herding family as well. The herder first gains knowledge about the horse's name, age, temperament, and brand mark. A herder looks at the age of the horse by examining its teeth. The favourable characteristics that are recognised in stallions are these: a wide forehead, wide nostrils, good teeth, thick hair and skin, long eyelashes and eyebrows, a big chest, slanting ribs, well-developed muscles, thick shins, sharp-edged hooves, and a loud squeal (Bayarsaikhan 2006). These characteristics are evidently for aesthetic reasons but have survival implications too, as thick hair and skin would be a benefit in the cold weather conditions, a big chest and well-developed muscles would mean the horse is faster, and the ability to vocalise loudly is an "honest signal" to indicate relative size and strength. In contrast to the stallion, good characteristics recognised in a mare are these: a big upper lip, high chest, round and broad body

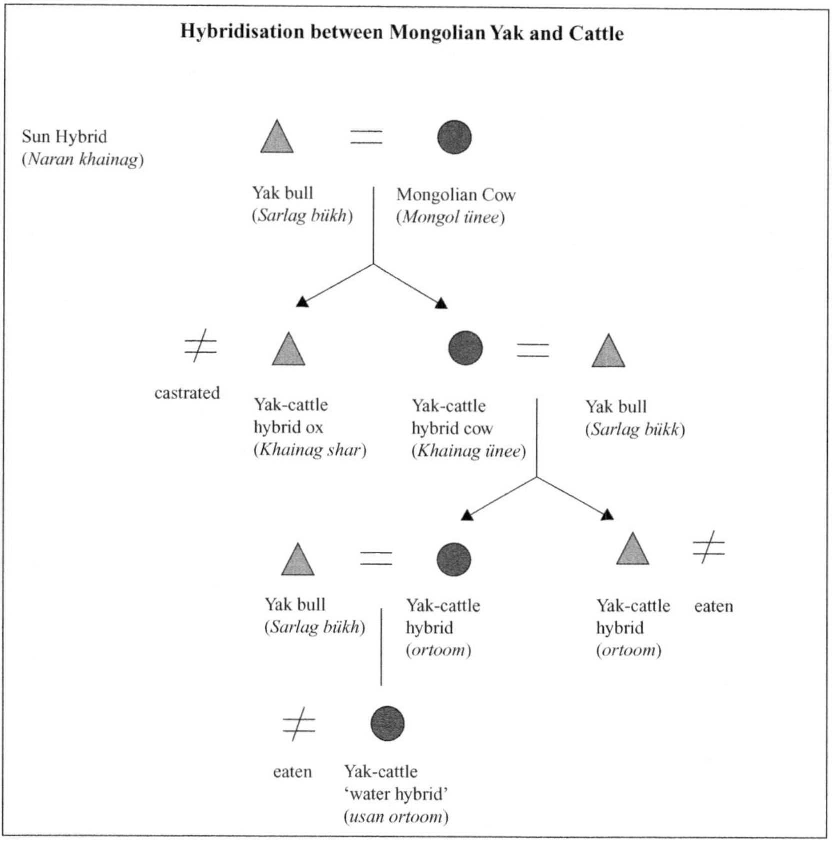

Figure 4.3. Mongolian cattle and *sarlag* hybridisation.

shape, a big stomach, straight, thickset legs, and thick hooves (Bayarsaikhan 2006). It is evident that these features are oriented towards the mare being able to carry a foal and give birth to offspring easily.

Ambling horses were used by nobility in the past and are still valued in Mongolia, as they are more comfortable to ride at a trot (see Chapter 5 for more details on ambling). Davaa, a neighbour to Dogsomjav and Boloro's brother-in-law, rode over to a wealthy herd encampment on the opposite side of the Selenge River and brought back a three-year-old palomino ambling stallion. This was an expensive exchange, at the cost of three young cattle. The stallion was chosen because he came from a different area, which meant that the horse herd would have an injection of new genetic stock. This beautiful young stallion almost immediately acquired two mares and one foal from another stallion's herd, already beginning a herd of his own within a few days after his arrival. In the case of Davaa's herd, he is likely to have ambling horses with an occasional individual exhibiting the unusual palomino coat colour as a result of inheritance from the stallion.

I asked Divangar, from the Mongolian State Agricultural University, what herders look for when selecting horses for breeding. Here is Divangar's reply, which was spoken in English:

> First, they are interested in lineage. Who are the parents? Second, they look at their physical features, size, or strength. Third, their behaviour is very important, especially the stallion, as it is important not to choose the wrong individual. Does it have good behaviour, quiet behaviour? The stallion manages the herd, so they look at herd behaviour. They [stallions] must look out for young ones, protect them from other herds and wolves, therefore behaviour is important.... If a stallion's behaviour isn't good, then all their foals will also be like this; therefore, Mongolian herders need quiet behaviour for their herd. The stallion's part is also more general [for instance, the stallion's coat colour]. Herders don't like black and white colours on the stallion. Some with a white head aren't good. If a young stallion has specific colours that aren't good then they won't choose them. If I have a stallion with a white head, or one with white eyes, then herders in my area don't like it. (Divangar, Ulaanbaatar, November 3, 2005)

Animals with white eyes are likely to be prone to blindness and those with white heads may be prone to sunburn and ultimately skin cancer, so there may be practical ecological reasons behind these coat colour choices. Even in ancient Pazyryk burials, the horses that were sacrificed were predominantly chestnut, brown, and less commonly bay, whereas the colours of dappled grey, mottled bay, skewbald, roan, and grey were absent (Rudenko 1970). As Fukui (1996) points out, the herders mentally record genealogy, mediated by coat colour.

PART II. NOMENCLATURE

Mongolians refer to co-domestic animals generally as "*mal*," encompassing the five kinds of herd animal (*tavan khoshuu mal*). In this part of the chapter I begin at the level of the herd, followed by identification according to each herding family, then down to the scale of identification of individual herd animals. Different species of livestock are formed into herds in Mongolia, with sheep and goats as one, and cattle (including yak) as another; the stallion and mares form one herd and usually the gelded male horses another (refer to Chapter 3).[3] The herds are also divided into "hot muzzle" (*khalun khamar*)

[3] Camels do not inhabit the Khangai Mountain area, where my two field areas were situated, so I was unfortunately not able to include camels in my research.

animals, referring to horses and sheep, and "cold muzzle" (*khuiten khamar*) animals, referring to cattle and goats. According to Humphrey with Onon (1996), these categories are based on whether the herd bunches together in cold weather or not, whether their meat provides "heat," and, symbolically, whether they are close or distant kin. Herders also classify pastures as "cool" (*seruun nutag*) or "hot" (*khaluun nutag*) grazing territories, which aligns with their classification of the herd animals. Hot-muzzled animals, for instance, are grazers like yaks and horses and are recognised by herders as being best suited to cool pastures located near rivers, on mountaintops, and north-facing slopes. Even food is divided into "hot" and "cold" energy foods: Mutton and horse are "hot" foods to be consumed in winter, whereas goat, butter, and milk products are "cold" foods to be consumed in the heat of summer.

All sheep and goats are herded together from one encampment to form a "sheep family" (*khonini khot*). Together, the sheep and goats are referred to as "*bog*"or just as "the sheep" (*khoni*), which is a simpler term than always referring to "the sheep and goats." Reference to the herd generically as "sheep" has a long history. A fourteenth-century text, the *Yuanshi*, rarely mentioned the term "goat," only "sheep" (Bold 2001). During the Socialist era of communal *negdels*, all animals were measured in terms of so-called sheep equivalent units, as a gauge of their value relative to one sheep (Open Society Forum 2004). Mongolian cattle (*Mongol*), yak (*sarlag*), and yak–cow hybrids (*khainag*) tend to be herded together and are all categorised under the generic term of "cattle" (*ükher*) by herders. According to Bold (2001), one of the reasons for this is that yak only live in the high Khangai and Altai mountain regions, so it is not included as a sixth species. The generic word for "horse" is *aduu*. Mongolians from an urban environment mistakenly refer to horses as *mor'*, but this is the more specific word for a gelding, or castrated male horse. *Azarga* is the word for stallion but is also used to refer to a horse herd, as one *azarga* is a numeric term for a horse herd consisting of between twenty and twenty-five individuals.

THE USE OF SYMBOLS

There is very little fencing throughout Mongolia, and a herd is generally free to roam and potentially intermingle with neighbouring herds. Animal rustling is prevalent, so animals are marked with a brand or a nick in the animal's ear, effectively showing which herding family the animal is aligned with. The sheep and goat herd and cattle herd are given earmarkings, or *im*, whereas horses (and sometimes camels) are given brands, or *tamga*. According to Vainshtein (1980), the use of the earmark is an ancient method of identification in Inner Asia, as the use of the *im* existed amongst early nomadic peoples in the Altai Mountains (see reference to the Pazyryk burials in Chapter 7). These methods of animal classification are also similar to other current herding cultures, from reindeer herding Eveny in Siberia (Vitebsky 2005); reindeer herding Tungus (Ingold 1974); cattle herding Maasai in Kenya (Galaty 1989); cattle herding Bodi in Ethiopia (Fukui 1979, 1996); to cattle farms in New Zealand.

The subsequent extract is a field note excerpt from Dogsomjav's encampment detailing how she chose sheep and goats from the herd to give to her youngest son upon marriage. She used earmarks to identify that the animals were part of her son's family's livestock and were no longer hers. Dogsomjav's youngest son had recently married a young woman from a neighbouring herd encampment:

> Ochero and Dogsomjav entered the sheep and goat pen with large shearing scissors. Dogsomjav selected out fifteen sheep and thirteen goats to give to her youngest son as his inheritance upon marriage. She identified each individual that was hers by the facial markings on each sheep or goat: a sheep that was brown around the eyes (*bor nüden*), a goat with white flecks on the ears (*buural chikhtei*), or one with a prominent white blaze down the middle of the face (*khaltar*). Ochero caught the selected individual and grasped it firmly about the head and shoulders, while Dogsomjav folded over the right ear and deftly snipped a small chunk from each side to form two half circles on either side of the ear. The goats made panicked bleating with alarming similarity to a child's cry and blood quickly soaked the white patches on their faces. When one cried out plaintively, Dogsomjav mumbled "poor thing" ("*khöörkhii*"). As we were walking back to the ger, Dogsomjav commented that her youngest son's wife was given thirty sheep for her inheritance, as her parents have a larger herd. She went on to say that she now has few animals left, as most have been given to nine of her children and she only has one more daughter to give animals to as inheritance when her daughter marries. Out of the horses, Dogsomjav is now left with only one mare and two geldings that she will continue to ride while herding. (Field note excerpt: Dogsomjav's autumn camp, October 19, 2005)

To designate that the sheep or goat is part of Dogsomjav's extended family encampment, the left ear has a cut (or tear) down the middle (*tsuulbar*), which is the symbol for the encampment as a whole. The mark on the right ear identifies individuals belonging to each nuclear family within the extended family, such as Ochero's family, Lhagva's family, and the new mark for Dogsomjav's youngest son's family. When members of the extended family look at an animal from their herd they can identify the individual animal not only on the basis of distinguishing characteristics but the particular individual or family to whom it is aligned. The structure of the sheep and goat herd therefore broadly mirrors the structure of the extended family encampment, with different family members associated with a representative number of animals.

The predominant mark that identifies a horse's association within a herding encampment is the brand mark, or *tamga*. Prior to the Soviet era the brand was an important symbol of social status and the branding iron (also called a *tamga*) was passed down through the generations from the patriarch of the encampment to the youngest son. In Caroline Humphrey's early work in Mongolia, she eloquently describes how

> ...the signs of the tamaga system were not simply addressed to a hypothetical stranger horseman riding through the steppe. They were also intended for the use of kinsmen in their relations with one another, and even, one might say, for an individual in his relation with his social role.... In Mongolia the land is wide and there are very few men. The numerous herds of horses become a reflection, almost a mirror image of the society of men, a magnified and recurring likeness of the social groups whose wealth they are. (Waddington-Humphrey 1974: 486)

The brand marks represent symbols that are important, or particularly auspicious, within Mongolian herding culture, such as these: fire, water droplets, the moon, a table, an animal pen, horse's bit, or camel's nose-peg (refer to Figure 4.4). Brands can function as protection for the herd; for example, the "fish" *tamga* signifies vigilance and should therefore protect the herd from wolf attack, and the "knot," which signifies eternity, should bring long life. During the Soviet era, the patrilineal system of *tamga* inheritance became less relevant, because individual families no longer had ownership of their animals; they were owned by the State. Each *negdel* had a *tamga*, which changed from the old style of symbols to incorporate Soviet iconography, such as the hammer and sickle, or were represented by initials in Russian Cyrillic. After privatisation of the livestock in Mongolia, herders again adopted more traditional-style *tamga*, often Buddhist in origin.

Before the Revolution (before 1921), it was illegal to have a brand of one's own choosing. My experience was that the use of the *tamga* is no longer as structured as it was in the past. Dogsomjav's encampment has a *tamga* iron that they use to brand the herd as an identification mark. The family's *tamga* was not inherited but was chosen by Dogsomjav's husband from a booklet that outlined the different types of symbol that could be used. Dogsomjav's family *tamga* depicts a Tibetan Buddhist astrological symbol in the form of a *chandmani*, or a "wish-granting jewel," represented by three circles above a crescent-shaped moon (*sara*). I asked Dogsomjav who had inherited the *tamga*, because her husband had recently passed away. She told me

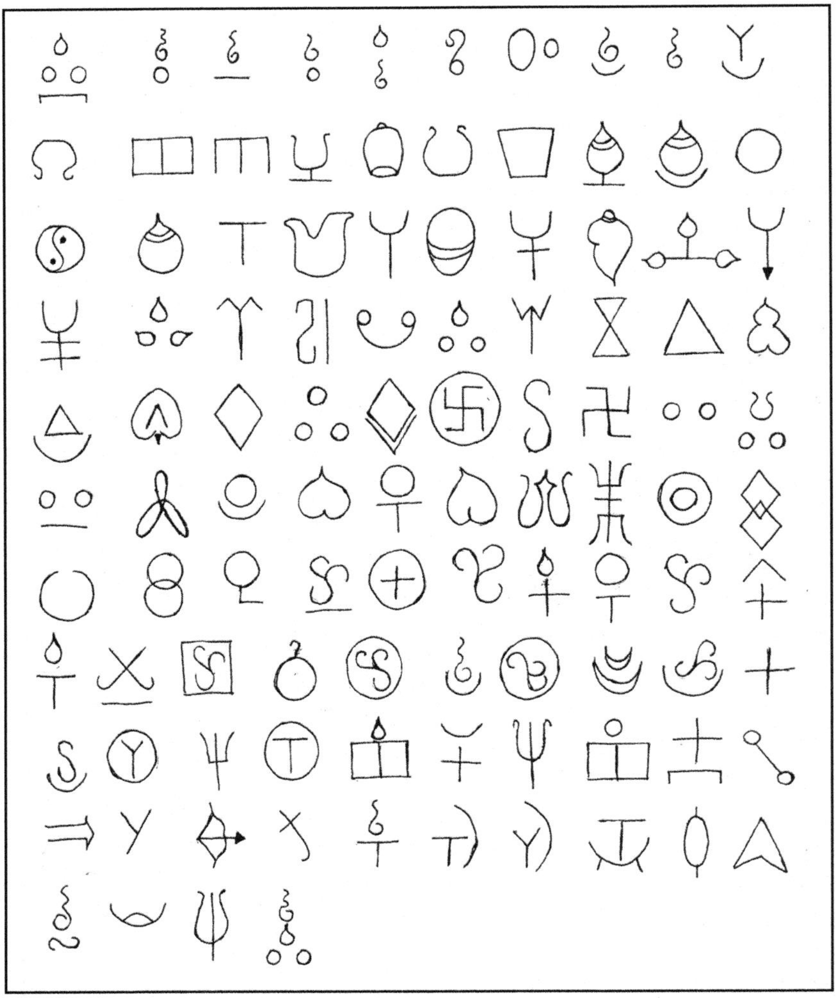

Figure 4.4. Drawings of *tamga* symbols. (*Note:* redrawn from a booklet on *tamga* symbols in Mongolia, author unknown.)

that Ochero, her second-eldest son, inherited the *tamga*, as her youngest son was living in the city and Ochero was the only son who remained in the encampment.

In the past, the symbols represented the social status of the family and there was a limit to the type of *tamga* a family could use. At present, horses may be exchanged between neighbouring families, resulting in one herd having a mixture of many different brands. Choijo's encampment does not use a *tamga* of their own, and I counted as many as eleven different *tamga* within Choijo's herd, where horses had been obtained from other herds with brand marks. Two neighbouring summer camps each had a stallion with the same *tamga*. When I enquired as to why the brands were the same, one of the women said that it was because the horses were originally from her father's herd, but because she married into the neighbouring encampment, she took the stallion with her family's *tamga* with her. I then asked how they knew which horses belonged to their own herd and the woman replied that they rely on their ability to recognise them individually by coat colour and markings. The *tamga* are no longer as accurate as they were in the past as markers of a herd's, or herder's, status, but the *tamga* is still a good means of identifying the history of an individual horse, particularly if the horse was exchanged or stolen from another herd.

INDIVIDUAL IDENTIFICATION

Mongolian herders identify individual animals by their external features, or *züs*. Based on their *züs*, a herder is able to recognise when animals are missing from the herd. Galaty (1989) found, in an analysis of Maasai pastoralists, that Maasai have a remarkable ability to remember the genealogy of the members of their herd, with symbolic categorisation of the herd animals contributing to an impressive spatial memory. The Bodi of East Africa (Fukui 1979, 1996) also have an intricate system of classification of cattle by colour and pattern, which is closely intertwined with their social organisation and even linked to homicidal warfare against neighbouring tribes (see also Evans-Pritchard 1940). Like East African pastoralists, Mongolians name their animals on the basis of colour, pattern, horn shape, markings, sex, and age. Mongolian herders have a remarkable ability to recognise individual animals, even from a young age. The herding boy, Öskö, at the age of six years (when I returned to the encampment in 2007) could identify each sheep and goat within the herd individually and select the young lambs and kids to match them with their mothers when it was time for nursing.

Herders in Dogsomjav and Choijo's encampments refer to animals in terms of physical appearance followed by their age group, whether they are male or female, or their breeding status as methods of identification. Age is especially relevant during milking. Calves are usually separated from second-year steers and heifers that are still feeding from their mothers in different pens. Individuals may be classified as age groups up to five years of age (see Table 4.2 for the different age groups according to herd animal). If it is any older than five years of age, then the individual is generally referred to by breeding status.[4]

[4] Mongolians include the period of gestation in the age of both people and animals.

Table 4.2. Categorisation of ages of four kinds of herd animal

Age	Horse	Cattle	Sheep	Goat
First year	*Unaga*	*Togal*	*Khurag*	*Ishig*
Second year	*Daaga*	*Byaruu*	*Tylög*	*Borlon*
Third year	*Shudlin*	*Guna, Gunjin*	*Shudlin*	*Shudlin*
Fourth year	*Khyazaalin*	*Dönö, Dönj*	*Khyazaalin*	*Khyazaalin*
Fifth year (adult)	*Soyoolon*	*Soyoolon*	*Soyoolon*	*Soyoolon*

Herd animals in Mongolia are identified in a different manner, depending upon whether they are horses, cattle, or sheep and goats. Identification tends to be primarily on the basis of facial and body markings in sheep and goats. Cattle (including *sarlag*) identification is broader and tends to be on the basis of age and sex, facial markings, or external features, like horns and height. Horses are mainly identified on the basis of overall body colouration, perhaps with additional markings for clarification. The names of the herd animals are unique to the type of animal in question. This is similar to the way Saami name their animals; according to Anderson (1986), dog names do not overlap with human names. They are selected from an inventory of terms used only for dogs, describing colour or texture of coat.

Sheep and Goats

When sheep and goats give birth it is particularly necessary for the herder to be able to identify which newborn lamb or kid belongs to a particular mother. Individual lambs or kids that are hard to distinguish are identified mainly by using coloured cloth (*khondog*); the herder then ties the same coloured cloth onto the mother's horn, or another part of the body. Different methods are used for identification: coloured cotton sewn into the ears; cloth tied around the leg or horn; or paint on the back or horns. Once the young are strong enough to find their mothers and feed for themselves, the coloured markings are removed. Both lambs and kids are given earmarks within the first few weeks after birth, or during castration (see the description in earlier text).

Beyond this key birthing time, a herder's contact with sheep and goats occurs mainly while out herding, so interaction is mainly on the level of the herd as a whole, rather than on a one-to-one basis. In most instances a sheep or goat is referred to on the basis of facial or body markings, such as white flecks on the ears (*Buural chicktei*), brown eye spots (*Bor nüden*), or a red collar-like mark on the neck (*Ulaan khüzüüt*). There are names, however, used to designate an animal's specific role within the herd. The leader of a herd (*Surgiin Manlai*) may be given the name of "*Darkhlakh*" (meaning

"Protector"), or if a member of the herd is particularly small and cute, "*Khenz*" (which literally means "Late Born").

Cattle

Cattle are identified in quite a different manner from sheep and goats, as women interact with individual cattle on a daily basis during milking for much of the year. In both Choijo's and Dogsomjav's encampments, each calf is named individually from birth. There are common names for the cattle that are not specific to one encampment; for example, the firstborn and lastborn calves of the year are usually called Firstborn (*Uuganaa*) and Lastborn (*Otgono*). Mongolian families sometimes also give these names to their firstborn and lastborn children. Other common names for calves are Multicoloured (*Tarlan*), Small (*Odis*), or Red (*Ulaanaa*). Some encampments do not give the calves individual names but refer to them by colour, so to a degree the naming process is dependent upon the families' own established naming system, or means of recognising individuals.

Calves are mainly named on the basis of differing physical features, for example, Hairy (*Bavainai*); Brown with Moon-Marking (*Sartai bor*); and Birthmark (*Menget*). In Choijo's encampment, because the majority of the herd consists of *sarlag*, the few individuals that are *Mongol* cows or *khainag* have their breed added to their name, such as Multicoloured-Hybrid (*Tarlan Khainag*); Dark Brown-Hybrid (*Khüren Khainag*); or White-Blue-Mongolian Cow (*Tsenkher Mongol*). This is an effective way of retaining knowledge of the genealogy of the herd. Calves are often given the same name as their mother, which makes it easier to match the cow with her offspring during milking time.[5]

Names are mainly based on physical characteristics but some cattle have names that describe emotive characteristics regarding the behavioural temperament of an individual, such as Moody (*Aasht*) or Bright (*Sergelen*). One example of a specific name for a cow with identifying characteristics that were unusual was a yak–cow hybrid from a neighbouring encampment who was particularly distinguishable in appearance. She had the face of a cow, when usually *khainag* faces resembled *sarlag* more than Mongolian cows, and her body was short and squat, with a particularly low-slung belly. The neighbouring family named her Nature (*Baigalaa*) after a woman from the nearest township with the same name. This woman had short legs and was overweight, so she resembled the strange-looking hybrid. We would chuckle during milking times as the comical-looking hybrid cow passed by us with the neighbouring herd.

[5] See Saikhanaa and the Calves, available at www.cambridge.org/9781107000902.

Horses

In Mongolia, the indigenous breed of Mongolian horse has a wide variation of colours, although the base colour is recognised by herders as bay (brown with a black mane and tail). In European horses, there are also specific descriptive terms for the different types of horse colour, such as palomino, piebald, skewbald, bay, and dun. These colour differences coincide with different breeds of horse from disparate geographic locations, yet Mongolian horses have the full range of coat colours. The description of a horse's coat colour can also be modified by adding adjectives for such distinctive characteristics as eye colour, baldness, colours on the mane or tail, or the horse's brand mark. White markings on the horse's fetlock or muzzle, or as a blaze down the forehead, can be used in combination with the horse's overall colouration. A white spot on the forehead is called "with a star" (*sartai*), or a larger white spot is "with a mirror-star," whereas a long white streak down the nose is "bald" (*khalzan*).

Bökönyi (1984) refers to an ethnographer who collected 321 expressions describing the colour of horses from Hungarian herdsmen, of which 12 referred just to colour variants of bay horses. I suspect this would have included all possible colour combinations. Vitebsky (2005) refers to an Eveny scholar who has compiled 1,500 Eveny words in reference to the colours and shapes of reindeer, but this number also includes body parts, harnesses, diseases, diet, and moods. Evans-Pritchard (1940) relates that there are "several hundred" colour combinations and permutations used by the Nuer to name their cattle. The *Lonely Planet Mongolia* guidebook states that "Mongolians have over 300 different words to describe the country's 3.1 million horses, mostly relating to colouring" (Mayhew 2001: 28). It is likely that this number of colours is an exaggeration, similar to the popular belief of an exaggerated number of Eskimo words for snow (Martin 1986). There is certainly a greater breadth of Mongolian descriptive terms for horse coat colour than I was able to record, but I endeavoured to list the main types. I noted herders referring to 78 different types of descriptive colour terms alone for Mongolian horses (see Table A.1 in the Appendix). The colours of horses can be divided into four main groups: white (*tsagaan*); black (*khar*); chestnut (*zeerd*); and light bay (*khongor*). Each of these four main categories can be divided into further subcategories. There are endless possible variations to these basic colour types by combining some of these descriptive colour terms. Such a wide range of descriptive terms gives an indication of the importance of the horse in Mongolian herding culture and the herder's remarkable ability to differentiate between individuals.

Herders can identify horses even when the coat colour is changeable. Some horse colours seem to be counterintuitive; for example, a horse that may have a white coat is referred to as "Brown," or *Bor* (although this corresponds with calling an essentially white horse "grey" in Western riding circles; see Hemmer 1990). A horse that is white as an adult could have been brown as a foal. A horse's coat colour changes during its lifetime or with the seasons, so calling the horse Brown is representative of a constant colour. Another reason for calling the horse Brown is because the Lord of the Heavens (*Tengger*) rides a white horse. A herder on earth should not assume a horse within the herd could contain such power so, to be humble, the herder refers to the horse as Brown (Bayarsaikhan 2006). The herder's skill in identifying individual horses is exemplified by a story that was recounted to me a number of times of a brown horse that was sent to a distant land but, after a long and arduous journey, managed to find his way back home. The herder identified that this was the same horse, even though the horse had evidently changed to the colour of white from the stress of the journey.

A few animals are what the herders refer to as "milk fed" (*ugjaa*) because either the mother was unable to feed her young (she may have been a young mother) or the mother died soon after birth. It is my impression that herders unintentionally develop bonds with these *ugjaa* individuals. In these cases, women in the encampment will give the young animal milk by hand and, as a result of this closer contact, the animal may be given a more descriptive name. One such case was a piebald gelding, who was a milk-fed orphan. He was given the name of Bambaruushka, a Russian word meaning Little Bear, because he was rotund as a foal. Bambaruushka was tamer than the other geldings in the herd and ridden by Dogsomjav's grandchildren because he had a particularly quiet temperament.[6]

Humans

People's names may change during their lifetime in Mongolia. Children are given temporary names at various stages. A child will not be given a name for the first month. Even when the baby has been named, the baby will not be referred to by his or her given name for the first few months but will instead be referred to by a baby name. Baby names, like animal names, may be based on colours, such as Yellow (*Sharaa*), or Red (*Ulaanaa*). These colours refer to the complexion of the baby, whether the skin colour is yellow or red in hue. When a child has a dirty face, he or she is described as being a "brown child" (*khaltar khuukhed*). The term *khaltar* is also used to describe the coat colour on animals; for example, a dark chestnut horse is a *khaltar mor'*.

6 For footage of Bambaruushka, see A Day in the Life, available at www.cambridge.org/9781107000902.

In other words, the same descriptive colour characteristics used in baby names, or nicknames, are used for both people and animals.

When I referred to Dogsomjav's three-month-old grandchild by name, Ochero corrected me by saying he's "not a person" (*khün bish*). People may actually be named "Not a Person" (*Khün Bish*) to keep evil spirits away. I was told that adult names are too "heavy" for a small baby. The baby is unburdened by the responsibilities usually expected of other members of a herding family, and at this stage the baby's survival is uncertain: There is a risk that the baby may die within the first few months. The baby is expected to fight for survival until it is strong enough to hold the burden of a "heavy" name. Applying a name means a higher degree of emotional attachment and that an individual has become a person. The gradations in naming may also be applied to the level of personhood in the herd animal as well.

Birthmarks are spiritually significant in Mongolia, particularly across generations of people within the same family. Mongolians are born with a mark or so-called Mongolian spot on their buttocks or lower backs. The birthmark looks similar to a bruise and usually disappears during childhood. These marks can be found on people throughout the extent of the former Mongol Empire, and many Mongolians trace them back to the lineage of Chinggis Khaan, a genetic but also spiritual link with their historic past. The importance of such marks is particularly interesting, as birthmarks are markers of reincarnation and rebirth in Mongolia. According to the Mongolian spiritual calendar, days of the week are aligned with a specific number and colour of mark, or birthmark (*menge*), for example, six white marks, or nine red marks (see Chapter 10 for more details on the Mongolian calendar). When an elderly person dies, the family members wait to see whether the next grandchild that is born is a reincarnation of their deceased grandparent. This is signified by the newborn baby having a similar birthmark to the grandparent, accompanied by a similar personality. The grandchild is effectively believed to be the reincarnation of his or her grandparent. Similarly, markings are important for the individual identification of the herd animals and herders value individual animals that have particularly distinctive marks; for example, Lhagva named a calf Menget, or Birthmark, because of the distinctive mark on her forehead. The spiritual significance, or auspicious meaning, behind bodily markings is not just restricted to humans but applies to the herd animals as well.

ANIMALS AS PERSONS

An individual animal may be recognised as a "person" in its own right, as recognition of individuality and particularity. As Mary Midgley points out, the origin of the word "person" does not equate with being human. It means a character or personage acted, one who plays or performs a part, a character. The word *persona*, in Latin, did not apply to slaves. In terms of the current law in the United States, the word "person" can apply to a corporation but not to a dolphin (Midgley 1985). Here, I am using the word "person" to apply to any being with recognised individuality, character, and personality.

The degree to which a name is applied to a specific individual, or refers to particular behavioural characteristics, depends upon the degree of contact the herder may have with an individual animal. Ungulates that are milk fed, or hand reared, may be given more pertinent names referring to the animals' character than those that blend in with the herd as a whole. It may be that Mongolian herders avoid using more human-like names for herd animals to refrain from developing close bonds, as ultimately it is necessary to slaughter these animals for food (see Chapter 10).

Similarly, babies are not given a proper name until they have passed a crucial stage when their chances of survival are greater and they have become socialised into the family. The herders do not view humans as being superior to all other animals; instead, each species is recognised as different and is therefore identified according to separate criteria. In other words, the names that animals are given are not anthropomorphic, as they are more descriptive of individual appearance and behaviour. Mongolian world view means that their "animacy create(s) a domain in which personhood transcend(s) the limits of purely human existence" (Humphrey with Onon 1996). Herders conceive of individual animals as having specific behavioural and physical characteristics, or as beings with an individual *personality*.

Barbara Smuts does not write of animals as persons, but she has experienced encounters with other animals with the presence of something resembling a "self." Until very recently, up until the past few generations, humans were generally very familiar with other non-human beings. It is Barbara Smuts's belief that all humans still have the capacity for engaging at an in-depth level with other beings, but few of us have the opportunity to do so in fast-paced, urban lifestyles. While conducting in-depth ethological research on baboons in Africa, Smuts (2001: 301) found that the baboons taught her to view all other beings differently:

> Before Africa, if I were walking in the woods and came across a squirrel, I would enjoy its presence, but I would experience it as a member of a class, "squirrel." Now, I experience every squirrel I encounter as a small, fuzzy-tailed, person-like creature. Even though I usually don't know this squirrel from another, I know that if I tried, I would, and that once I did, this squirrel would reveal itself as an utterly unique being, different in temperament and behaviour

> from every other squirrel in the world.... My awareness of the individuality of all beings, and of the capacity of at least some beings to respond to the individuality in me, transforms the world into a universe replete with opportunities to develop personal relationships of all kinds.

In relation to animistic cosmology, Harvey (2006a) describes "persons" as those with whom other persons interact with varying degrees of reciprocity. One of the main reasons why I describe Mongolian herders' outlook as animist is that they engage with herd animals as persons. They name each animal individually, ascribe personality and specific characteristics to different individuals, and treat them as beings with emotion, with whom they can communicate on a social level. As Scott (1996: 73) describes in relation to the perspective of the Cree in Canada, "All persons engage in a reciprocally communicative reality." Anderson (2000) writes of the Siberian Evenki's view of the world as an "interagentivity," in which animal and human persons attend to one another. In our society some humans may be treated as objects and animals may be treated as a means of production for human gain, "but just as people are not always persons, persons are not always people" (Knight 2005: 2).

My premise is that if an individual is given a name, especially one based on unique characteristics, then the animal is likely to be treated with more respect and recognised as an individual, even to the extent of viewing the individual with greater empathy. Phillips (1994) found that cats and dogs in a laboratory are not generally named and are treated differently from the cat or dog in the home. A scientist's avoidance of naming laboratory animals reveals a connection between naming animals and the potential to develop emotional ties. Phillips (1994: 121) states that by naming animals "we socially construct individuals and create a narrative account of the meaning of their lives." Once an animal or human is classified as an individual entity, then that individual is no longer just one of a mass, or just a member of a herd. In reference to Western farming, it has become difficult to treat each member of a herd as an individual. Juliet Clutton-Brock (1994: 34) writes this:

> It is inevitable that once the number of animals owned becomes large, say in the thousands, their individual identities are lost. This has been the inevitable result of industrial farming in the modern world where the vast number of domestic animals are required to feed the ever-increasing human population ... all bred to look identical and reared in rows of cages, to be harvested when required.

CONCLUSION

The way people refer to animals is related to their relative importance within society and the level of interaction or engagement with the animals as individual actors. Mongolians have multiple scales for identifying their animals, on the basis of the species, the herd, or on the level of the individual. This individual level of classification is crucial to understanding the herders' relationship with their animals. Through a system of categorisation according to coat colour, pattern, age, sex, and other distinguishing physical or behavioural characteristics, each member is an entity in its own right and not just one of a mass of objects within a herd. In an egalitarian manner, each animal is attributed with specific behavioural and physical characteristics, unique to each being. Mongolians do not view the herd animals purely as economic products to be exploited for human consumption, or as mere things: They are primarily thought of as individuals with a personality, essentially as "persons" engaged in a reciprocal relationship.

Nor is there just one type of selection occurring (either artificial or natural); there are multiple layers. Selection of the animals is influenced by factors such as climate, predation (both by wolves and humans), and illness, but also selection of the next generation on the basis of the animal's reproductive success and the quality of both morphological and behavioural traits. In Mongolia, the selection of coat colour is not only an important means of individual identification but is an important factor in the selection of a breeding male. This human part of the selection process is referred to by Mongolians as the "folk" (*arat*) or the "people's method," as a result of skills in selection and breeding developed over the centuries.

Nevertheless, herders do not *make* livestock; they assist in providing the right conditions to allow the animals to *grow*. Animals are not made, as a cart or dairy products are made; they have their own development that is beyond human control. Tim Ingold (2000: 87) insists that domestication should be referred to in terms of "growing" rather than "making," because the animals are nurtured just as children are, through growth into adulthood: "The forms that all these creatures take are neither given in advance or imposed from above, but emerge within the context of their mutual involvement in a single, continuous field of relationships."

5

MULTISPECIES ENCULTURATION

The most remarkable feature of this kind of interspecies communication between such different kinds of being is that it takes place at all. (Patton 2003: 88)

We are not different from other species by having a culture which they lack; we are different in that our culture, like our shape, is different from theirs. (Rose 1984: 46)

INTRODUCTION

We expect domestic animals to interact with us and to understand our communication with them, even though we are separate species and our evolutionary paths are quite different. Humans and ungulates are able to respond to cues and gestures that are a product of their social gregariousness, engaging in both intraspecies and interspecies communication. Seminal papers on animal communication by Dawkins and Krebs in the late 1970s recognised that animals communicate between species and are even deceitful in their intentions towards one another, particularly between predator and prey (Dawkins 1978; Dawkins & Krebs 1979; Guilford & Dawkins 1991). Dogs, humans, and sheep are able to assist one another in a farming situation through one species communicating *signals* and another species *receiving* those signals and reacting to them. As Donna Haraway (2007) points out, the whole process would not work if sheep did not know how to understand dogs and if dogs did not know how to interpret them. In a Western context, cross-species communication between a farmer and a sheep dog may develop into a wide array of voice commands, whistles, and pointing signals. As Tani states (1996: 389), "human beings can control the behaviour of domesticated animals by means of vocal signs taught in the context of a bond of intimacy between human and animal."

In this chapter I suggest that both herder and herd animal respond and engage with one another in a multitude of subtle ways:

> Nasaa caught a horse from amongst the horse herd as a fresh mount for the day's sheep and goat herding. He had been using the lasso-pole as a whip (*shelbuur*) to direct the sheep and goat herd and then attached the cord to the end of the pole to form a loop, or lasso (*uurga*), to catch a horse from the herd of geldings. He rode up to the geldings, selected out the horse he was to catch, nudged the horse he was riding to make it rush forward rapidly, and then inverted the lasso just above the horse's head. The young herder said "brrr, brrr," meaning "stop, stop," to the horse even before he had the lasso over the horse's head. The caught horse halted immediately when he felt the pressure of the tightened lasso on his neck and remained still as Nasaa dismounted. Nasaa removed the saddle and bridle from the sweating mount he was just riding and released him to graze with the rest of the herd. He then placed the same bridle and saddle on the fresh horse and rode off to tend the sheep and goat herd, still grazing nearby. (Field note excerpt: Dogsomjav's spring encampment, May 12, 2005)

This extract from my field notes exemplifies the type of communication that occurs between herder and herd animal, as two species that engage in a mutual understanding with one another. I noticed that Mongolian horses are generally calm with external stimuli while being ridden, such as the lasso-pole extending out beyond the horse's head within its peripheral vision. The herder communicates to the horse that is being ridden, mainly through body language, but the herder also vocalises to the goats, sheep, and, in a removed sense, to potential wolves to warn them away from the flock through yells and whistles: This is a complex system of multispecies communication. The riding horse even recognises that there are some signals that are to be responded to, but other signals that are to be ignored. Lestel et al. refer to this hybrid community of shepherd-dog-wolf-herd as a "balanced system." These are "communities in which interspecific communication has a privileged role and whose social organisation leads to changes on *both* sides" (2006: 161). This chapter and the accompanying visual material illustrate how sounds and body language are crucial aspects of cross-species engagement and enculturation.[1]

[1] See Vocalisations, available at www.cambridge.org/9781107000902. The accompanying video segment could be thought of as a visual and auditory dictionary or ethogram of herder–herd animal signals. Intertitles detail the meaning of each vocalisation, followed by a visual representation of the herd animal's response to the sound the herder makes.

MEANS OF COMMUNICATION

Ungulates use a variety of modes of communication involving different senses, such as visual displays, physical contact, vocalisations, and social odours. Many animals identify one another or assess others' physiological and emotional state on the basis of smell. Stallions defecate in a particular area to define their home range, and they sniff the dung left by other stallions or geldings. This is a signal left in the landscape that not only defines a sense of ownership of a place, but leaves a mark where the herd has previously been. Dung piles left by stallions are a signal to other competing male horses to mark paternity rights to females (King & Gurnall 2007). A passing horse will be able to establish, from this visual and scent-based signpost, to which herd the dung belongs (perhaps similar in function to the brand mark or *tamga* for herders; see Chapter 4).

As humans, we do not have the same ability to communicate as adeptly through our sense of smell but we are able to communicate within a similar sensory range by the means of sight and hearing. This means that, by learning their habits and behaviour, humans are able to communicate with ungulates. Mongolian herders are aware of the herd animals' differing levels of sensory perception and recognise that some of it is beyond the realm of humans. They recognise the herd animals' sensitivity to their environment, particularly the horse's strong homing instinct and sensitivity to external stimuli. Here is a common explanation, this time given by Bayarsaikhan (2006: 105), of the Mongolian horse's remarkable sense of direction and navigation skills:

> In the darkness of night, a rider pulls the lead of his horse to stop it, after which the horse holds its breath sensing the surrounding environment. Then, it points to a [specific] direction with its ears and if one lets the horse go by itself [one] arrives at a camp among grazing animals. When there is marsh, ravine, or other difficulties on the road, the horse suddenly stops, tilts its head and moves [off] finding a convenient way all by itself.

Mongolian herders described the herd animals as being "sensitive" on a number of occasions. I interpreted this to be in reference to the herd animals' emotional responsiveness and sensory acuity. The following is one such example of the herders' identification with the herd animals as emotionally responsive beings. A young, black bull came to visit the cattle herd at Dogsomjav's autumn site from a neighbouring encampment a few kilometres up the valley. A few days previously Ochero had killed a two-year-old steer or *byaruu* because he was literally "crazy" (*zerleg*); the *byaruu* had developed a tendency to compulsively turn in tight circles. Because a steer is quite large, he was knocked on the head and then slaughtered a short distance away from the *ger* behind the winter shelter. Some of the steer's blood had spilt in the long grass as the carcass was being processed but the remains were no longer visible, as most of the blood had soaked into the ground.

> A bull arrived from the neighbouring encampment and sniffed around where the *byaruu* had been slaughtered. The bull was bellowing and panting with his mouth agape, obviously in an agitated state. When I asked Dogsomjav what the bull was doing, she said "he is crying for the *byaruu*." (October 22, 2005)

> The black bull returned again today in the evening and started bellowing at the same place. Most of the cattle were over at the autumn milking area but became very agitated with the commotion the black bull was causing. The oxen joined in the fray as the younger steers and a few of the cows started chasing and butting each other. They kept disrupting the cows that were hobbled by butting them. A group of cattle raced over to investigate where the bull had been bellowing, sniffing at the ground where the crazy *byaruu* had been killed. They became quite agitated, butting each other. Lhagva asked me to drive them away, because "otherwise they will begin crying over the dead *byaruu*." They raced around kicking their legs in the air defiantly when I drove them away. (Field note excerpt: Dogsomjav's autumn encampment, October 23, 2005)

Dogsomjav later explained to me that the bull's name is Bukhandai and that he visited because he was born at their encampment. She said, "his mother is Khar Bedere and his grandmother is Ulaanaa" (see Chapter 4, Figure 4.2). I was amazed at this behaviour, at the sensitivity of the bull to the smell of the *byaruu*'s blood and even that he remembered not only his place of birth but also the members of the herd that he had grown up amongst. Dogsomjav and Lhagva did not seem surprised at all and quite readily attributed the bull's behaviour to the emotion of sadness for another individual's death and that the bull was exhibiting this emotion by "crying."

When I began working with the cows during milking, Lhagva instructed me on some behavioural aspects of the cows. "They don't like strong scents, like perfume, and cows don't tend to like bright colours and they prefer the one-piece *deel* [jacket]." One reason for this is that work *deel* are rarely washed, so they become impregnated with the scent of the family's herd.

The cows, particularly *sarlag*, inhale deeply when handled and immediately attempt to flee, or give a deep, low grunt, when they cannot recognise a person's scent. After the first few milking occasions, the *sarlag* became quite calm in my presence. I was told, when I first began riding the horses in Mongolia, that Mongolian horses do not like people wearing bright yellow, or red clothing, or wearing hair loose about the head. Similar to a Spanish bull in a ring responding to a red cape, horses are also responsive to bright colours. They have dichromatic colour vision and respond to differences in luminescence, which means they can discriminate between colours such as red and blue (Macuda & Timney 1999). Herders have a good knowledge of the herd animals' sensory perceptions, such as sensitivity to colour, smell, and movement, and they adjust their behaviour around the herd animals accordingly.

THE EMOTIVE USE OF SONG

The Sheep Song (Mongolian folk song)	*Toig (Khalkh ardin duu)*
My pure white sheep	*Khundan tsagaan khon min*
Why do you neglect your lamb?	*Khuurga youndaa goloo be?*
The smell of your milk	*Suunii chin uner*
On its tail!	*Suulend n baina shuu dee!*
Release, release, release [your milk]!	*Khos, khos, khos, khos!*
The spring will come	*Khavrin tsag bolno doo*
The snow will melt	*Khajuugiin thas khanzarna daa*
Dry winds will blow	*Khavsarga salkhi garna daa*
Who will stay beside you?	*Khajuud chin you chin khevtekh be dee?*
Feed, feed, feed, feed [your baby]!	*Toig, toig, toig, toig!*
The warm weather will come	*Urin tsag irne dee*
The grass will grow	*Uulin nogoo urgana daa*
Your udder is full of milk	*Uurag suu chin chinene dee*
Who will come to greet you?	*Urdaas chin you chin ugtakh be dee?*
Feed, feed, feed, feed [your baby]!	*Toig, toig, toig, toig!*
Abundant summer will come	*Delger zun bolno doo*
The grass will fill the earth	*Delkhiin nogoo urgana daa*
Your udder is full of milk	*Delen khökh chin chinene dee*
Who will suck from your udder?	*Tuuniig you chin khökhokh be dee?*
Feed, feed, feed, feed [your baby]!	*Toig, toig, toig, toig!*

Women and children sing to mother sheep, encouraging them to release milk from their udders for their newborn lambs, repeatedly singing "*toig, toig*" to them. Herders not only sing to their animals but refer to the animals as "singing" themselves. There is a name for the pandemonium of noise before lambs find their mothers after sheep are released from the pens each morning, referred to as "*khoinni nair*," or the sheep chorus. This chorus is very loud for a couple of minutes and then suddenly dies down once lambs find their mothers and begin to feed. In other words, the herders sing to mother sheep and speak of the sheep singing themselves. The following is a field note excerpt from spring, when many of the young animals require assistance feeding. The herding family are engaged every morning and evening, matching mother sheep and goats with their young:

> At this time of year [spring] lambs and kids are born day or night and may require assistance from the herding family in order to feed. Newborn lambs and kids that are more capable of feeding from their mothers are housed in a small wooden shelter with a central fireplace for warmth. Weaker lambs and kids that need assistance to feed are housed in one corner of the smallest family ger within the herd encampment. Nanny goats are brought inside at dawn for milking and to feed their kids. Upon entry into the enclosed space it is pandemonium, as the nannies and their young call to one another plaintively, while women encourage the goats to release a small amount of milk by singing "*zu, zu*" [release, release] to the goats. Other family members attempt to match mothers with their offspring and hold the goats still to allow the kids to feed. When the nanny goat is reunited with her kid, she may turn her head around to sniff the kid before releasing milk. If the nanny isn't responsive towards her young and won't release her milk, Saikhanaa holds the kid with its tiny body propped upon his knee. If the newborn is still unsuccessful at feeding itself, milk is poured into an ox horn with a teat on the narrow end, to feed the kid by hand. Some nannies seem to readily accept any kid feeding from them, just as they accept any human milking them, while others take a bit of coaxing. After feeding their young, the goats are pushed out the door to graze for the day. (Field note excerpt: Choijo's winter encampment, March 15, 2005)

Herding women sing to the female animals to encourage them to release milk to their offspring, "to make the animal regain its spirit, be touched, and develop good character" (Bayarsaikhan 2006: 271). Singing to the individual animal is accompanied by rubbing or scratching just above the udder to calm the mother. To encourage nanny goats to feed their young by releasing their milk from their udder, herders repeatedly sing "*zu*"; to ewes they sing "*toig*"; to horses "*guurii*"; to Mongolian cows they sing "*ööv*" (ow); and to *sarlag* they sing "*khos*" or "*suug*." One reason for the different sounds vocalised to Mongolian cows and *sarlag* may be that the two species vocalise quite differently; cows low but *sarlag* grunt (hence the domestic yak's taxonomic name of *Bos grunniens*). The use of the word "*ööv*" is reminiscent of the sound a cow makes when lowing, or "mooing," whereas the word *khos* is closer to the grunt of a *sarlag*.

Before I went to Mongolia, I was told by a Mongolian couple that there are herding legends about female camels shedding tears when they finally accept their young. If a mother camel does not accept her young, the members of the herding family sing and play a particular musical instrument (the *morin khuur*) and specific camel songs to the camel to encourage her to accept her newborn calf. As a response to the music, the camel may relent and accept her rejected calf.[2] A respected professor from the Mongolian State Agricultural University, who specialises in Mongolian camels, informed me that camels are particularly emotional and have been known to shed tears when the mother loses her calf. This professor told me that a Scandinavian explorer,[3] trekking through the Gobi desert, felt that he had to leave a camel behind because it was too weak and slow, but for two weeks the camel shed tears and he could not bring himself to part with the camel. When the explorer finally left he cried himself. As Tumerjav (2003: 102) suggests in relation to Mongolian camels, "As they are sensitive and willing animals, it is believed that camels are able to love, cry and sense joy."[4] Clearly, Mongolians believe that ungulates are behaviourally complex enough to respond emotionally to music and song.

Many Mongolian songs feature the beauty, strength, and power of the Mongolian horse and its capabilities. A popular Mongolian folk song, entitled "Mother's Love," was screened as a music clip on the local television station during the time I lived in Ulaanbaatar. This song was adapted from an old Mongolian legend, which exists in various versions. The music clip featured an extended family on the expansive grassland steppe. Tuul described the legend in the following way:

> A woman loses her husband and in her heartache she rejects her newborn baby (one version is that the baby is born with dark-coloured skin). Her wise old father takes the rejected baby and places it out on the grassland steppe. Their horse herd gallop towards the bundle but refrain from trampling on the baby. The mother then recognises her error and runs to the baby in alarm, accepting the baby as her own. (Field note excerpt: Tuul, Ulaanbaatar City, February 23, 2005)

This story is interesting in that Mongolians recognise the horses as having enough insight to avoid trampling on the small baby. The legend could be based on practical knowledge of horse herd behaviour. An added detail of

[2] *The Weeping Camel* (Byambasuren & Luigi 2004) is a feature-length film based on this remarkable story. Refer to The Interlude for the legend of the *morin khuur*, or horse-head fiddle. To Mongolians the *morin khuur* is a powerful object in its ability to evince emotion in persons (including camels) who hear it played.

[3] This was likely to be Sven Hedin's exhibition from 1926 to 1935.

[4] Camels are able to secrete tears from the eye as an adaptation to the dry desert environment (Gauthier-Pilters & Dagg 1981).

this story is that the wise old father had seen that horses avoid trampling on the eggs of a small, brown bird that nests in the grass; therefore he knew that the horses would avoid trampling on the defenceless baby. The theme of a mother's bond with her baby seems to be particularly important within Mongolian herding culture, both in terms of the herders themselves and the herd animals (see Chapter 6 for mother–offspring bonds). The herders' attentiveness to vulnerability and dependence leads them to recognise other animals as having the capability of occupying the role of protectors and nurturers, as other beings are part of the extended family.

While out herding, or travelling over long distances, riders sing traditional long-songs that are directed towards the horse they are riding. The horse responds to the herder's singing by stepping in a "more energetic manner." When I would accompany Dogsomjav or other members of the family herding the sheep and goats, they would sing songs. Mongolian herders have a large repertoire of songs, particularly about love but also numerous songs about horses, herding, the land, and the mountains. Erdenekhuyag jokingly proposed that, "if there was a competition to single out the best and most varied songs regarding horses around the world, Mongolia would probably win."[5]

Song is important during the training and racing of racehorses. Prior to any training session, or race, while they circle clockwise around the training posts the young jockeys sing "*giingoolokh*," elongating their vowels and chorusing together. Erdenekhuyag told me, "when young horses hear *giingoolokh* they are attracted to the source of the sound and prick up their ears" and "when the horses hear this singing they release their bowels in preparation for racing." Both horse and rider are able to prepare mentally for racing, as the sound of the singing and the action of circling around the training posts is consistently followed by the act of galloping over long distances. In a famous Mongolian film, entitled *Tod Magnai,*[6] an old horse weeps when he hears the sound of the young jockeys singing "*Giingoolokh*" and witnesses the younger horses preparing to race.

The prominent use of song in Mongolian herders' relationships with their animals exemplifies the use of vocal signals to elicit not only a desired response but to elicit an emotive response from the animal. Mongolians relate how singing is "a sound that touches the heart of the animal." In Western academic traditions, particularly in fields such as ethology or comparative psychology, there is a reluctance to attribute emotive responses to

[5] I interviewed Erdenekhuyag in Ulaanbaatar as he had recently written a book about horses in Mongolia, *The Horses of Chinggis Khaan* (2004).

[6] This means "Clear Forehead," a reference to the winning racehorse (see the Praise Ceremony in Chapter 7 for another reference to this term).

animals, particularly from a source as abstract as song or music. The fear of applying emotions to animals has not always been a sensitive subject, as Charles Darwin had no qualms in stating that animals feel pleasure, pain, happiness, and misery (Darwin 1868). Gradually, scientific evidence is mounting that animals can express forms of emotion and are therefore sentient beings, if not fully conscious (Dawkins 2006; Dennett 1995, 1996). Ingold (2000) makes the persuasive point that non-human animal awareness is becoming a legitimate topic but that it is still based upon a mind–body dualism because the research is phrased in terms of animal "minds," which makes the assumption that consciousness must reside purely in the mind, when in Ingold's opinion it is essentially an extension of the body and the surrounding "landscape." Similar to Anderson's (2000) use of the term "sentient ecology," I prefer to equate such emotive responses as "sentience," rather than the more problematic term of "consciousness." In Mongolian herding society, sentience means that humans and herd animals have the ability to express feeling and emotion to one another; engage in cross-species communication; understand their place within herd social hierarchy; and to possess a knowledge of belonging (or acceptance of individuals as part of the social group).

HERDING LANGUAGE

Mongolian herders use a wide range of verbal signals and commands to communicate with herd animals, both on an individual level and on the level of the herd as a whole. I noted five general sounds commonly used for all herd animals; thirty used for sheep and goats; and sixteen for cattle and seventeen for horses. These vocal sounds also convey different meanings depending upon context, repetition, elongation of vowels, tone, and loudness.

Sheep herding in Western society involves a wide range of whistles and commands, but these are directed towards the sheep dog to round up sheep. Mongolian herders do not use dogs to assist with herding but call directly to the sheep and goats themselves. Upon hearing the herders signalling, I was surprised that the herd responded to the wide array of commands and whistles. The calls not only function to direct the herd while out on the grassland but also to distract wolves from attacking the flock. A herder is effectively announcing the presence of potential human threat to the wolf to deter an attack. This involves multispecies communication, as the herder is signalling to the sheep and goat herd, as well as any wolves that may be nearby.

Herders use different calls depending upon whether the herd is moving out to pasture or whether they are more settled and grazing. A common signal used by the herders when the herd is moving out to pasture to encourage the herd to move forward at a faster pace is "*khai*." The alteration of the pitch, tone, or loudness lends different meanings to a signal, as this same call of "*khai*" in a different tone while the herd is grazing means for the herd to come back, or to change direction. Short, sharp sounds are meant more as commands; longer sounds are intended to be reassuring and convey to the herd that everything is fine; and yet other sounds are almost pleading, meaning "please obey." Words directed towards the animals are sometimes onomatopoeic in that the signal is very similar to sounds the animals make themselves. Goats snort through their nostrils as a warning of potential danger to the rest of the herd. Herders vibrate through the lips with a forceful "*ppp*" sound, which is directed towards kids to stop them from chewing on the canvas of the *ger* or the laundry hanging from the line. This "*ppp*" sound is remarkably similar to a goat's warning signal.

Herders indicate the direction towards which sheep and goats should be progressing, whether they should stop and graze in the same area, or move on to nearby pastures. The sheep and goats do not have the ability to respond to the level of complexity of recognising left or right, but if the herd is heading in the wrong direction the call is repeated until the herd responds to the vocal signal correctly. Dogsomjav, for instance, would direct "*chat*" towards the sheep, signalling "don't go there" when the herd was moving towards the forest. If the herd was a long distance away, Dogsomjav would call an elongated "*chaaai*," indicating for the herd to change direction. Members of Dogsomjav's family use vocalisations while out herding that are specific to their encampment. Other vocalisations are specific to the region, yet some even extend up into Tuva in Russia (Vainshtein 1980). At the Bulgan field site the herders say "*zelee, zelee*" to encourage calves to return to their pen, whereas herders use "*güürgee, güürgee*" in the same context at the Arkhangai field site (this is also because it is a signal towards *sarlag* calves rather than *Mongol* calves).

When the herd is moving from one location to another, the herder will use body language in combination with vocalisations to communicate intention. Children use any object that is at hand, such as dried dung, or sticks, to throw at the animals. Herders on horseback generally use a *shelbuur*, which is a two-metre-long birch pole with cord attached to the end, effectively functioning as an elongated whip (this leather cord can be looped on the end to form an *uurga* for catching horses). I noticed the difference in having the *shelbuur* compared to only using my voice while herding, as individuals would respond to urging forward from the cracking sound of the *shelbuur* accompanied by the signal of "*khai*" with more efficacy than if I used the vocalisation alone, probably as a result of the urgency of the sound from the whip. In the past, herders would use a *duuguur*, a hide sling and stone, for herding *sarlag*. When yaks heard the sound of a stone whizzing past their heads, the yaks would move in the opposite direction to the sound. The use of tools, such as throwing dung, or a sweeping movement with the *shelbuur*, acts as an extension of the body to convey the herder's intentions to the animal.

The widest range of vocalisations, at the level of the herd, is used with the sheep and goat herd (*bog*). Herders do not tend to let the sheep and goat herd wander far, or to allow them to venture within tracts of forest where wolf attack is more likely. The horse and cattle herds are monitored far less by herders and it is only for specific reasons, such as milking, that the herd is directed back to the encampment and into a corral, as horses and cattle (particularly *sarlag*) are more adept at defending themselves against wolf attack.[7] Usually the horse and cattle herds are left to fend for themselves and are free to wander independently once they are out to pasture. A smaller range of vocalisations are directed towards cattle and horses at the herd level and these signals tend to be more generalised in form, such as an undulating "ooh" sound, or a range of whistles.

Enculturation can be defined as "the process whereby individuals learn their group's culture, through experience, observation and instruction."[8] Although the term was originally used in relation to humans, recently the concept of enculturation has been applied to great apes, like chimpanzees and orangutans, and even dogs, horses, and goats, in their ability to read human cues. Enculturation tends to be more effective in behaviourally plastic animals that have early associations and a dependence on humans (McKinley & Sambrook 2000). Michael Tomasello used the term of "ontogenetic enculturation" to explain why human-raised great apes are particularly successful in cognitive tasks (Tomasello & Call 2004). The term "phylogenetic enculturation" (as opposed to ontogenetic enculturation) has been used to explain why dogs respond to human signals like pointing, or merely gazing with the eyes, as cues to locate hidden food or toys when wolves are unable to respond successfully to such visual cues (Hare et al. 2002). In other words, there has been genetic selection for dogs to look towards human faces in times of uncertainty.

Kaminski et al. (2005) suggest that domestic goats also have cognitive social skills that have developed as part of the domestication process. In experiments they found that goats follow the gaze of other goats and most notably they also followed humans when they pointed to a hidden food

[7] On a number of occasions I witnessed *sarlag* and *khainag* bunching together and then actively pursuing dogs in defence of the rest of the herd.

[8] This definition, retrieved November 13, 2007, is from the American Psychological Association: enculturation (n.d.). *Dictionary.com Unabridged* (v. 1.1), available at http://dictionary.reference.com/browse/enculturation.

source (although their responses were not as successful as those of dogs). There is also anecdotal evidence that sheep react to cues of the shepherd without the need for herding dogs. This is clearly evident in the Mongolian herding situation, where goats and sheep are directed by vocal commands, rather than through commands to a dog, who then rounds up the herd. It is my view that this indicates that both sheep and goats have an ability to read and respond to human communicative cues and, through enculturation, sheep and goats adjust to the social behaviour of the herder.

COMMUNICATING WITH INDIVIDUALS

When the Runa of Amazonia talk to their dogs, Kohn (2007) describes the language as "transspecies pidgin," with a reduced grammatical structure and the incorporation of elements of communication from both human and canine realms. Mongolian herders also use a kind of transspecies pidgin towards the ungulates they herd. This section focuses on vocalisations that are directed specifically towards the individual animal rather than the herd as a whole. The sounds are based on the Mongolian language but the words are specific to the ungulate species towards which it is addressed, subtly nuanced with a variety of inflections and tones to convey meaning. A vocal signal is elongated or shortened, depending on the pace required: A short sharp "*chu*" to a horse means to break into a fast pace, whereas "*chauuu*" means to move into a more leisurely trot. Vocalisations are often repeated in quick succession for emphasis; the use of the signal "*chu*" is represented by "*chu, chu.*" The vocalisation of "*chu*" (meaning "go"), accompanied by a whack on the rump with a short whip (*tashuur*), is much more likely to initiate an appropriate response than the use of the vocal command alone. The extensive use of the word "*chu*" is so prevalent in use amongst Mongolian herders that it is even extended beyond the context of a command towards a horse to include an exhalation when lifting heavy objects, or when using physical exertion during wrestling bouts.

Vocalisations towards herd animals vary not only depending on the social context, whether the signal is directed towards the herd or to the individual, but also depending on the gender and age of the animal. "*Güürgee*" is used to indicate to young animals, such as Mongolian calves, to go inside the pen, whereas words like "*drrr*," "*suu*," and "*tut*" tend to be used for trained oxen. If a mare has the propensity to kick out during milking, she is hobbled on three legs. The mare cooperates in response to the vocal command of "*bogus*" by holding her leg up to assist the herder.

At milking times, both horses and cattle react to common vocal commands with the correct behavioural responses a large proportion of the time

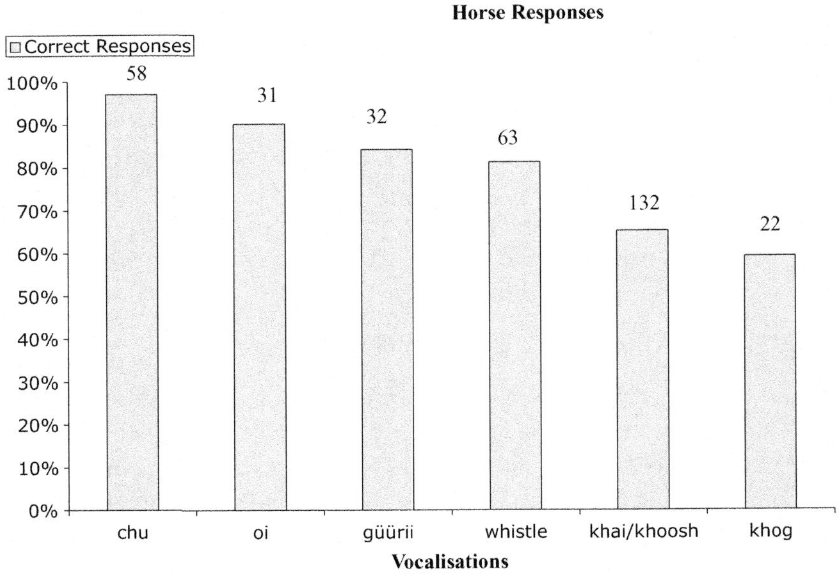

Figure 5.1. Proportions of correct horse responses to common herder vocalisations during milking times.

(see Figures 5.1 and 5.2).[9,10] It is evident from the histograms shown in Figures 5.1 and 5.2 that both horses and cattle respond better to an encouragement to move forward at a faster pace, for example, "*chu*" for horses or "*khöög*" for cattle, than commands to stop, such as "*khog*" for horses or "*drrr*" for cattle, or reprimands such as "*khai*" and "*khoosh*." The vocalisations of "*khai*" and "*khoosh*" were often used for horses to prevent them from moving away or kicking while being milked ($N = 132$), whereas "*oow*" (*ööv*) and "*kha*" were often used to encourage the cows to stay still and calm during milking ($N = 108$).

9 Here is a brief explanation of the meaning of vocalisations for Figures 5.1 and 5.2. Horses: "*chu*," move forward faster; "*oi*," hey!; "*güürii*," release milk; "whistle," stay still; "*khai/khoosh*," stop that!; "*khog*," stop. Cattle: "*khöög*," move forward faster; "*khudg*," come here or get up; "*zel/güürgii*," go to pen; "*oow/kha*," stop moving and stay calm; "*drrr*," stop; "*khai/khoosh*," stop that! Note that I have combined some vocalisations to simplify categories when the sounds have similar meaning (e.g., "*oow*" and "*kha*").

10 At both the Arkhangai and Bulgan field locations, during milking times I recorded when herders used a common vocal command toward an individual animal; whether the individual's response was correct or incorrect; and whether the vocalisation was directed towards a young herd animal, an adult male, or an adult female. I did not record repeated vocalisations to the same individual animal during a milking session and only recorded the behaviour if it was clearly in response to a command (within five seconds that the vocalisation was made).

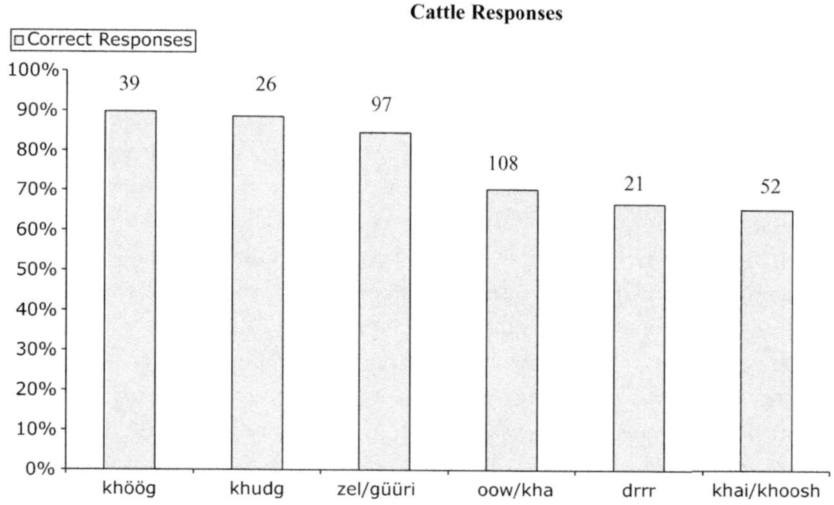

Figure 5.2. Proportions of correct cattle responses to common herder vocalisations during milking times.

A herder will call an animal by name to gain cooperation, or a specific action, from a particular individual. During milking, calves are called to the gate of the pen before they feed from their mothers. Calves are readily able to identify their own names during feeding times, perhaps because they are required to learn the sound of their mother's call in order to feed. After the first few weeks, the newborn calves learn to respond to their individual names by coming to the gate of the calf pen when it is time to be fed. Cows that are regularly milked respond to their names and the women sometimes accompany a specific command by announcing the cow's name (e.g., "Short-tail, stay still," or "*Oodon, khai*"). Oodon was responsive to both her own name and the name of her calf. If I called her calf's name, saying "*Menget,*" then she would vocalise to her calf as well. It is evident that both cattle and herders can identify others on an individual level, particularly in relation to milking, which is due to the strong mother–offspring bond. Herders use this ability of the herd animals to recognise one another on an individual vocal level to successfully communicate to specific individuals within the herd.

A combination of both body language and vocalisations is likely to be the most effective strategy for communicating on an individual level. Body language is mainly directed towards an individual herd animal, as it is more difficult to express intentions corporeally to a number of individuals within a herd. The activity of milking is a context that requires both vocal and bodily signals to elicit the amount of cooperation required for milking by hand. A female with a full udder can choose whether to release milk or not.

The herding women encourage ewes to release their milk by exclaiming "*khos*" and slapping the udder with the palm of the hand, or encourage a mare by uttering a high pitched "*güürii*," alternated with sucking noises similar to the sucking sound made by a foal (the foal is also kept beside the mare during milking and fed briefly before and after milking). When hobbling cows, members of Choijo's family use different signals from Dogsomjav's family, because one family has predominantly *sarlag* and the other *Mongol* cows. Saikhanaa approaches a *sarlag* to hobble her for milking by saying "*khaa*," refraining from physically touching the *sarlag* but holding the tie used for hobbling in an outstretched palm. Perhaps this gesture to the *sarlag* has a similar function to the human wave, a non-threatening gesture, indicating that the tie is not a menacing object, like a stick. This technique is quite different from the one Lhagva uses to stop *Mongol* cows from walking around before she hobbles them: She places a hand gently on the rump and says "*oow oow*" (*ööv ööv*).

Lott and Hart (1977) examined the socialisation of cattle by Fulani herders in sub-Saharan Africa; they formulated a link between the Fulani's dominance towards cattle to aggression and dominance in their social relations with other people. Lott and Hart illustrated how the Fulani maintain dominance over cattle by their response to "broadside" threats in bulls. If a bull turns at right angles towards a herder, the herder recognises this as a threatening gesture and responds by a counter threat gesture with a raised stick, or a yell, combined with a quick charge towards the bull. If the bull does not respond with a submissive signal, he will be hit with the stick. I observed similar communication between the Mongolian herders and the herd animals, where defensive signals by the herder were due to an inappropriate behavioural response by the herd animal and this was counteracted by the herder through the combination of a verbal and a physical command. Mongolian herders respond to threat gestures, such as a broadside threat from an ox, by raising the hand, or a stick, and this is accompanied by a yelled vocalisation, such as "*khoosh*." Occasionally a stick or dried dung is thrown towards the threatening individual.

Herders spend large parts of their lives socially interacting and communicating with the herd animals, whether it is a herder spending most of the day out on horseback tending the sheep and goats, or while milking the various species of animal throughout the long days of summer. An efficient and effective way of communicating with the animals has developed over time through specific verbal sounds and bodily movements. From birth, both herder and herd animal learn to convey and to read these gestures. Sometimes the intentions may be misinterpreted and result in threatening or defensive responses, but in most circumstances a mutual understanding exists in which both comply with one another's expectations. As young persons, both herder and herd animal become enculturated into herd society by means of interspecies communication, in the form of verbal and bodily signals. As Despret (2004: 115) so eloquently describes,

> Talented riders behave and move like horses ... mere thought from one may simultaneously induce the other to move. Human bodies have been transformed by and into a horse's body. Who influences and who is influenced, in this story, are questions that can no longer receive a clear answer. Both, human and horse, are cause and effect of each other's movements. Both induce and are induced, affect and are affected. Both embody each other's mind.

THE EMBODIMENT OF HORSE AND RIDER

A common story regarding the intelligence of animals is the story of Clever Hans, a four-year-old horse that, in 1907, could apparently solve complex mathematical problems, spell, and even discriminate between colours and tones of music by tapping his right foot on the floor to provide a correct answer. A psychologist, Oskar Pfungst, discovered that there was no deception on the part of the horse's questioners, or for that matter, on the part of the horse. Remarkably, the horse could answer questions correctly from complete strangers without any prior training. Pfungst eventually deduced that the horse was in fact reading subtle cues, as Clever Hans could not answer if he could not see the person asking the question, or if the questioners did not know the correct answer themselves. The questioners would provide cues by leaning forward slightly to observe Clever Hans as he tapped his foot, and once he reached the correct answer they would relax and stand in an upright position. The horse was responding to subtle body positions that even the questioners were not aware of. This famous case has been labeled the "Clever Hans fallacy" because the horse did not understand the mathematical problems put to him. Nevertheless, as both Vinciane Despret (2004) and Vicky Hearne (2000) point out, in his ability to switch from kinesthetic cues to visual cues, Clever Hans was indeed clever, as he had the ability to read the body language of his human questioners, cues that were so subtle they were beyond human consciousness.[11] Despret (2006: 116) suggests that both Hans and his questioners were actively influencing one another:

> In addition to the human bodies involved who were deeply engaged and interested, Hans's feats also testify to his capacity to be actively engaged in the game

[11] For an example of the subtle cues and body language involved in professional dressage, see http://www.youtube.com/watch?v=zKQgTiqhPbw, retrieved September 6, 2010.

> proposed, to give intense attention to minimal gestures expressing human desires, expectations and affects, and to respond to them in a remarkable way.

The communication between Mongolian herder and horse are just as complex, intuitive, and subtle. I became attuned to riding the Mongolian horse, and thus learnt how to signal and communicate my intentions in a cross-cultural, cross-species situation. My previous experience with riding horses was in New Zealand when I was growing up. The circumstances in Mongolia were very different. The general temperament of the horse, the method of communicating with the horse, and the style of riding were all unlike my previous experiences. The style of riding in Mongolia is more closely aligned to Western riding, rather than British riding techniques, presumably because the function of both Mongolian and Western styles is based on the effective herding of livestock, whereas British riding was originally based on European military techniques. It was necessary for me to learn the signals and cues the Mongolian horse had been taught, while the horse had to adjust to the slight differences in my methods of direction. To engage with the horses as a Mongolian herder would, I consciously attempted to unlearn some methods that I had been taught as a child. I had to adjust, for example, to remain in the saddle while trotting, rather than rising to the trot. I found, however, that each horse I rode adjusted quickly and we came to a mutual understanding of our methods of communication. The old herder, Dalkhaa, said to me in a somewhat surprised tone, after he observed me briskly cantering on horseback over to his encampment, "you ride like a Mongolian!"

In approaching a horse on foot in Mongolia, the rider always approaches from the left-hand side (*zuv tal*). The established method of mounting a horse throughout Central Asia is from the left side, and this is the same technique as in British riding practice. The horse is also approached from the left-hand side to be milked (although Mongolians always milk cows and *sarlag* on the right-hand side and sheep and goats from the back). The British technique is to groom a horse every day and to train it to accept handling on any part of the body, patting the horse while riding if it responds to a difficult command correctly. Mongolian horses do not appear to require the same amount of bonding or physical contact, as they are only ridden for short periods of time; instead the horses socialise and bond with other horses within the herd, grooming and leaning their heads on one another. Most of the Mongolian horses I encountered were at ease with being ridden, if sporadically, because they were caught and tethered as foals for mare milking, yet they did not respond well to pats, grooming, or being fed by hand.

The engagement between Mongolian herder and horse is quite different on the ground, when a horse is amongst the rest of the herd, compared to when the herder is directing and signalling to the horse from its back. Herders catch horses while the whole herd surrounds them within a corral, or milk the horses surrounded by mares and their young. The herders are very aware of the horses' defensive behaviours and the potential of being kicked by a particularly threatening horse. A herder watches the horse's body position during milking times and particularly the flickering of a horse's ears to gauge the horse's behavioural responses. If the horse makes a threat gesture, such as flattening the ears against the head, the herder yells "*khai!*" to the horse, meaning "stop that!" This is essentially warning the horse not to escalate its response into a kick or bite.[12] The men vocalise loudly while handling horses within a herd, especially when horses kick one another (a frequent occurrence with so many horses confined within a small area, such as a corral). Yelling functions to announce the herder's presence, so that the horses know where the herder is located, but herders also hold the view that vocalising loudly makes the horse less wary.

In New Zealand as a child, I established a bond between one particular horse and rode only that horse on a daily basis for a number of years. My horse was contained within a small paddock, on his own or with one other horse, but not integrated within a herd. The horse seldom had contact with other horses and was expected to bond with me, the rider. Mongolian horses engage mainly with other horses and are only ridden as one individual selected from a whole herd of horses. Each horse may be ridden for only a few days at a time and then left to roam freely for months at a time (although the frequency of use depends upon the size of the herd). When a horse is ridden, the horse takes on a different status, leading the herd as a whole, whereas any horse would usually behave according to its own social standing within the herd (see Chapter 6).

Horse Gaits

Mongolian herders have a large vocabulary for recognising many different types of gait. I noted forty-two descriptive terms used by Mongolian herders for different types of horse movement, divided into five broad categories of gait: walk (*alkhakh*), amble (or pace, *joroolokh*), trot (*khatirakh*), canter (*shogshokh*), and gallop (*davkhikh*). According to Bökönyi (1984), Western experts recognise twelve different gaits of the horse but these go back essentially to four main types: walk, trot, canter and gallop (ambling is usually categorised as a kind of trot). Herders tend predominantly to trot while herding or travelling any distance. Mongolian herders mainly walk or trot horses

[12] The words "*khai*" and "*khoosh*" are used amongst all herd animals in Mongolia and are even used by herders when reprimanding children.

during the colder months to allow the horse to conserve energy, but in the warmer months the usual gait is a jogging trot over long distances (*khatirakh*) and a more animated trot over shorter distances (*ergöö*). Faster gaits, such as cantering or galloping, are reserved for catching animals or racing, and they are mainly used during the warmer months when the horses will not tire easily.

Specific gaits have developed in particular "horse–human" societies and one such movement is the "amble" (*joroo*) in Mongolia. An "ambler" horse uses a gait that differs from a trot, and this results in the ride being smoother and more comfortable while the human is seated in the saddle. A horse usually trots with opposing legs striking the ground, whereas an ambler will strike the ground on one side, then the other in succession, essentially replacing a trotting gait with a pacing gait. Mongolians recognise different kinds of amble based on movements analogous to other familiar animals or speeds, such as the foal amble, sheep amble, camel amble, slow amble, water-like pure-bred fast amble, high amble, low amble, and the forced amble. Most horses cannot amble at all and are limited to a trot. Those that can are highly valued in Mongolia for their "smooth" (*zöölön*) gait and are selected from the herd as a riding horse for this purpose. Bayarsaikhan (2006: 148) refers to a particular kind of ambling horse as a "water pace horse" (*usan yaavdaltai mor'*), describing these horses as being "respected and loved by their owners. Herders ride such horses in celebrations and special events for [their] beautiful look." This wide vocabulary for horse movement is indicative of the awareness of herders to the subtleties of body movement and behaviour when riding a horse.

There are many Mongolian techniques and traditions in relation to riding a horse, and here I outline just a few. A herder mounts a horse as gently as possible, as if stepping into the saddle. Getting on a horse smoothly, even if the horse begins moving away before the rider sits down in the saddle, is the sign of a good rider. The rider directs the horse by the means of "neck-reining," in which the reins are held in the left hand, while other items, such as the *uurga*, *shelbuur*, or *telshuur*, are held in the right hand. When riding, a herder does not tend to direct a horse with his or her legs, preferring to use the reins, position of the body, vocal commands, and, if unresponsive, a whip on the hindquarters. In fact, the legs and heels of the rider do not contact the sides of the horse because the Mongolian saddle has large flaps to protect the stomach and side of the horse from the heavy stirrups. Herders don't tend to vocalise to the horses much while riding, except for the generic command of "*chu*" ("go"); instead, commands are predominantly through body language in the form of the rider's position in the saddle and signals using the reins. The stirrups are short, which means the seated rider is more comfortable when trotting with the horse over long distances, a necessity while herding all day, or on long journeys.

Mongolians do not rise to the trot but sit down in the saddle. In Ann Game's view, sitting in the saddle while trotting means that a rider is more at one with the horse. She states, "it is at sitting trot that I most live the centaur, that I feel most with KP [her horse] and most alive. In sitting trot, the rider has much more contact with the horse than in rising trot. ... Unless you allow for the connection, you cannot do it" (Game 2001: 9). Because the saddle is hard and it is tiring to sit in the saddle all day while out herding, the rider rests by repositioning in the saddle to one side or another. The rider may also stand up in the stirrups to rest over long distances, or stand up for rapid sprints when chasing after livestock on horseback. When a herder finishes riding a horse, the saddle is not taken off immediately until the horse has cooled down. After the horse is cool, the herder partially lifts the saddle a number of times and sets it back down before removing it completely to let the air on the horse's back dry the hair gradually.

CONCLUSION

Mongolian herders do not tend to "talk" to the herd animals, as people converse with pets in Western society. The herd animals are not routinely groomed, caressed, or spoken to like small children. The herd animals are free to engage with individuals from their own species, as well as other species throughout their lives. An individual animal is spoken to in a specific manner, similar to what Kohn (2007) calls a "transspecies pidgin": a mixture of the sounds the animal itself makes and sounds from Mongolian as a language. Mongolian herders have species-specific vocabulary that varies depending on whether the animal is young or old, male or female, or whether it is directed towards the herd as a whole or to an individual. Vocalisations are also dependent upon the circumstance, for instance, whether a horse is being ridden, castrated, herded, and even when a herder is encouraging a horse to drink. In this sense, Mongolians do not tend to anthropomorphise with regard to the herd animals because they do not speak to animals as if they are humans; instead, herders are able to engage with what they know an individual animal will respond to and understand.

Agustin Fuentes suggests that this sharing of human–animal biology and evolutionary history could be best described by the notion of "shared personhood": This is the "recognition of a shared interpretation of and response to interactions with environmental and social stimuli caused by common physiological and related biological systems" (Fuentes 2006: 126). This understanding of an animal's sensory perception extends to body language, which is particularly exemplified when a horse is ridden, as the horse is expected to respond to cues and signals through the body, while ignoring others. I propose that co-domestic ungulates in Mongolia have been subject to *phylogenetic enculturation* through their social interaction with herders over numerous generations, developing the ability to read signals through the use of vocalisations and body language. These signals are species specific and context dependent, resulting in a complex array of social cues through both voice and corporeal communication towards multiple species of social mammal.

AUTUMN

Plate 17. Ulaanaa (Red) ox with a water cart.

Plate 18. Training a young ox (shown on the right).

Plate 19. Dogsomjav setting up the autumn encampment.

Plate 20. Horse hitching post and washing line.

Plate 21. Sacred *Övöö* with stallion skulls.

Plate 22. Human out herding with horse, sheep, and goats.

Plate 23. Mongolian *sarlag* bull (*Tarlan bukh*).

Plate 24. Ochero dismounting after the day's herding.

6

TAMENESS AND CONTROL

What does that mean – "tame"?" … "It is an act too often neglected," said the fox. "It means to establish ties." "'To establish ties'?" "Just that," said the fox. "To me, you are still nothing more than a little boy that is just like a hundred thousand other little boys. And I have no need of you. And you, on your part, have no need of me. To you, I am nothing more than a fox like a hundred thousand other foxes. But if you tame me, then we shall need each other. To me, you will be unique in all the world. To you I shall be unique in all the world." (*The Little Prince,* Antoine de Saint Exupéry 1943).

INTRODUCTION

The subject of animal domestication brings out a complex mix of moralities: those of care and control; training and manipulation; domination and subjugation; and mastery and paternalism. Within this chapter, my aim is to unravel where Mongolian herders are situated in relation to their interactions with the herd animals amongst whom they live. The purpose of this chapter is also to draw out the key factors that contribute to the herder–herd animal co-domestic relationship. In Chapter 3, I detailed how the members of the herding family herd the animals in accordance with the behavioural inclinations of the herd animals. The herding family does, however, introduce some important elements that assist them in taming and controlling the herd animals' behaviour, such as organising them into multispecies herds, separating the young from their mothers, obtaining milk from the female animals, and castrating the males.

A key behavioural feature of a co-domestic animal is *tameness*. The process of taming is a phenomenon that is acquired within an individual's lifetime, but some animals have a greater genetic predisposition for tameness than others. In the Introduction, I detailed how the behavioural trait of tameness has been recognised as an important aspect in the domestication process. This was eloquently demonstrated by a fox farm experiment whereby foxes were selected on the basis of their tameness, not only resulting in behavioural changes in the foxes' degree of friendliness and lack of fear of humans but also resulting in physiological changes commonly found in domestic animals, such as floppy ears, variation in coat colour and pattern, and even patterns of reproduction.

According to Juliet Clutton-Brock, "A tame animal differs from a wild one in that it is dependent on man and will stay close to him of its own free will" (Clutton-Brock 1987: 12). Mongolian herders attempt to socialise young animals to become tame and therefore more manageable later in life. They view co-domestic animals as those animals that are nurtured by and engage with humans, yet individual animals are also described as having different behavioural traits, such as being "tame," which essentially means that the animal is calm and placid (*nomkhon*). Mongolian herders derive the word for "to tame" from this behavioural characteristic (*mal nomkhruulakh*). This chapter describes how specific factors encourage tameness and addresses how herders manage to attain control over the herd animals, despite the fact that the herds roam freely and are not held in a captive environment.

PART I. TAMENESS

The quality of tameness is one of a wide range of characteristics that constitute animal personalities. Mongolian herders comment upon whether a particular animal is "tame" or "quiet" (*nomkhon*), "bad tempered" or "moody" (*aashtai*), "lazy" (*zalkhuu*), "violent" (*dogshin*), or "calm"' and "peaceful" (*dölgöön*). Each individual is recognised as having particular behavioural characteristics, or the equivalent of a personality. A favoured behavioural characteristic is for the herd animal to exhibit placidity and tameness (or to be *nomkhon*). An analogy could be drawn with a Mongolian child, who may also be described behaviourally as "quiet" (*nomkhon*) or "wild" (*zerleg*). Just as children are nurtured to behave in a socially correct manner within their human social group, so too are the young herd animals nurtured by the herders to behave in a socially correct manner. I observed that a Mongolian herding family uses similar attitudes towards teaching both human children and young herd animals, not so much through punishment and reward but in allowing them to learn through trial and error and example.

When I first began helping with the cow milking, Lhagva warned me about the traits of one or two of the cows. *Khar Bedere* (Mottled-Black) was threatening and "violent" (*dogshin*); this cow would swing her horns at anyone that came near, apart from Lhagva. Divangar commented to me, "It is very interesting that some animals inherit wrong or bad temperaments.

Some animals are just bad tempered. Usually, if the family continues to use their animals and the animals adapt to that family, then they are very calm."[1] Herders describe particular horses as being "quiet" (*nomkhon*) or "moody" (*aashtai*), preferring to use calmer individuals for herding purposes (refer to Glossary IV for behavioural terms). They know which horses are easy to catch yet are difficult to ride. I was saddling a strawberry roan gelding and Dogsomjav warned me that he might be difficult while I saddled him. She was concerned because she knew that he was likely to move away and kick out while being saddled, and the horse behaved just as she predicted he would.

LEADERS AND ORPHANS

Those animals that are more attached to humans are often orphans (*unchen*) and have been milk fed from birth. If a young animal loses its mother, or the mother is incapable of feeding her own offspring (because the mother may be particularly young to be giving birth), then the herding women assist the young animal by feeding it milk by hand. Lambs and kids are fed with a rubber teat attached to an ox horn, or from a bowl generally used by the herding family. The lamb or kid quickly learns to rely on humans and will approach whenever someone is nearby, expecting food. These young are initially kept inside the families' own home, along with others that are not strong enough to nurse from their mothers without assistance from the herding family. Once the milk-fed orphan is capable of grazing on grass, it is encouraged to socialise with other lambs and kids and ultimately graze like any other individual within the herd. When young are more capable of feeding for themselves, they are still strongly bonded to members of the herding family but are repeatedly repelled if they approach any member of the family; they are encouraged to join the rest of the herd, a similar process to being weaned, although herders wean these individuals earlier than a nanny or ewe would. This means that the orphans are not too reliant on the herding family and do not become inappropriately bonded to humans instead of associating with other sheep and goats.

A male goat, Shargaa (meaning "Yellow"), was a milk-fed goat in Dogsomjav's encampment. As an adult he was larger than most of the other goats and considerably more attached to the herding family. While out herding, especially when he was hungry in the spring, he would approach us and graze nearby. The young boy, Öskö, would even place a racing saddle on his back

[1] Interview with Divangar, Mongolian State Agricultural University, Ulaanbaatar, November 3, 2005.

and ride him around the pen. The family pointed out to me that Shargaa's human attachment was unfortunate because it meant that killing him for food would be considerably harder because of their mutual attachment to him. Herders like the animals to be quiet and placid (*nomkhon*) but not too closely bonded to humans and reliant on them for survival. It is important for the herd animals to socialise and bond with other members of the herd but just as important for the herd animals to accept humans as members of their social group.

Firstborn animals of each season are held in high esteem and are given the name of Uuganaa, meaning Firstborn. A firstborn is usually one that is born at the start of spring (when it is still well below zero degrees). Upon the birth of a newborn calf, an offering of distilled vodka (*nermel*) is made to the Buddhist shrine. The human blesses the calf by applying three marks of *nermel* upon the forehead of the newborn, while saying to the calf, "may you be a well-rounded cow and lead 10,000 cattle." The birth of the first female horse of the season is celebrated with extra food to summon the birth of the livestock for the year and plenty of dairy products. At the start of the mare-milking season, the firstborn foal is caught with a lasso-pole (*uurga*), with a blue silk sash (*khadag*) tied on the end. This foal is tied at the first place along the tethering line and is "respected and loved during the whole summer" (Bayarsaikhan 2006: 218). These firstborn individuals are tamer than other young in their age cohort because they are required to spend a longer period of time within the family *ger* when early spring conditions are still extremely cold, with highly fluctuating weather conditions.

Lead animals, often goats, are crucial to the responsiveness of the herd as a whole. Within the sheep and goat herd, a few individuals direct the rest of the flock, usually lead goats. In Dogsomjav's herd, two castrated male goats (*serkh*) guide the rest of the flock. The lead animals are not always castrated males, however, and may be older females. Mongolian herders have specific names for these animals: *Tolgoilogch*, meaning "Leader," or *Daguulach*, meaning "Causes Others to Follow." Herders coordinate with these lead animals by calling for these individuals to "change direction," and then the rest of the herd follows their lead.

Tani (1989, 1996) describes how different pastoralists throughout the Mediterranean and Middle East have lead, or guide, sheep (and goats). There are guide-wethers, female leaders, and breeding females or males that are selected as the lead animal within different regions. They assist the herder by taking the lead when crossing rivers or walking through snow. All are hand-reared and specifically trained to take the lead position in the flock and respond to vocal signals from the herder. Tani's view is that herders in Central Asia, including Mongolia and Tibet, do not have the equivalent of these lead animals. My research does not align with Tani's comments, as herders do identify leaders and these lead animals still act as intermediaries between the herder and the flock; however, in the case of the Mongolian *Tolgoilogch*, the training is not specific to them as individuals. The training process is reliant upon the general responsiveness and inclination of the animal itself, whether it is a naturally dominant member of the herd, rather than through purposeful training of the leader when young (refer to Chapter 2 for a discussion on Tani's view on the role of the sacred animal in Mongolia). The herder's approach in Mongolia is to nurture the herd as a whole, which requires an egalitarian outlook, without favouritism or treating the animal as the equivalent of a pet. Likewise, the attitude within Mongolian herding society is to take care of everyone within the herding community, not just singling out individuals for special treatment. Nonetheless, contingencies such as extreme weather conditions, parental survival, and other factors do require that some animals have differential treatment from others. An animal's own personal history is reflected in the animal's later personality and role in the herding encampment.

THE SIGNIFICANCE OF MILK

Ingold (1980) would categorise the type of pastoralism Mongolians engage in as "milch pastoralism" because the social ungulates are important for the nutrition their milk provides. He refers to reindeer pastoralism that relies primarily on the meat of the reindeer in contrast as "carnivorous pastoralism." There are reindeer herders who live in the Taiga forest to the north of Mongolia who engage in milch-type pastoralism and are recognised as having close relations with their reindeer. Care for the reindeer is entrusted to women and children, who milk the reindeer. Each reindeer is named and comes to its name when called. The reindeer are essentially treated as members of the family (Aikio 1989; Carruthers 1913). These reindeer of the Tungus are much tamer than reindeer further north because of the close human–reindeer contact while being milked (Ingold 1974, 1980). This close bond between the Tungus and the reindeer and the link to early reindeer domestication alerted me to the importance of milking herd animals.

Milk is indeed of utmost importance in Mongolian pastoral society, not only nutritionally but as a strong mechanism for the continuing co-domestic relationship. Milking, rather than the more distant and less interactive herding, is a crucial part of the one-to-one interaction necessary for an animal to remain tame. Perhaps it would be more accurate if the herders' role were commonly described as "milkers" rather than "herders." Milking is crucial for both the link between the nursing mother and dependent young and between the female animal as the provider of milk and the human requiring milk for daily sustenance. If the animals were not milked, then the herd animals would not be particularly reliant on humans at all, as Mongolian native breeds do not rely on humans to be fed and are generally capable of surviving on pasture alone. The main reason why male animals do not become uncontrollable is that most are castrated before adulthood and are handled regularly during the early, formative stages of development during the milking process.

A female herd animal can choose to let down her milk or not, so the herders use conditioning signals to get the milk flowing, particularly the action of the calf sucking from the teats; in the case of milking mares, the herder mimics the sucking noises of a foal. Using young animals to stimulate the flow of milk is also practiced with bovines in Tibet, Africa, and India (Amoroso & Jewell 1963), but using this method to obtain milk from mares may be unique to Inner Asia. To obtain milk from cows the young calf is induced to nurse briefly, triggering the flow of milk from the udder. A young herder removes the calf and then one of the women milks until the flow of milk ceases. It is not possible to draw all the milk down from the fine ducts and usually a small amount remains in the udder. The young animal is then allowed to nurse again, extracting the remaining milk that the mother chooses to let down for her offspring.

A calm and quiet person milking will extract more milk from a cow than a loud extrovert. Having a strange person present can also be inhibiting. Cows, particularly *sarlag*, are sensitive to any new people during milking, because they are not part of the cow's recognition of the milker as part of her "herd," which includes both cattle and human members. I generally assisted with hobbling the cows and handling the calves but on the few occasions that I attempted milking, the herders would suggest a *Mongol* cow, rather than the more temperamental *sarlag* cow. I was unable to extract much milk, even though I would attempt to milk as quickly and deftly as possible, as the cow was more reluctant to let her milk down for anyone but the usual woman who milked her.

Surrogate Mothers

The establishment of the parent–offspring bond is crucial to Mongolian herders, because they cannot milk feed or supplementary feed many animals by hand at one time. In spring, a great deal of time and energy is spent getting young to feed effectively from their mothers and to gain enough strength to survive through the spring. If a cow loses her calf, a surrogate calf may be used to keep the milk flowing, so that the herding family can continue to milk the cow throughout the milking season, which may mean that a calf could feed from two or even three "mothers." Likewise, if a cow cannot feed her calf, a surrogate cow will replace her to provide milk.

Herders are aware that the initial recognition between a cow and her offspring is through the keen sense of smell. For a cow to accept another calf, the milking women place the hide of the dead calf over the surrogate to attract the cow to the smell of the hide. The cow then remains occupied in sniffing and licking this hide while being milked. The women stroke the surrogate cow and calm her by singing "*oow, oow*" (*ööv*) if it is a Mongolian cow or "*suug, suug*" if it is a *sarlag*, while the strange calf is feeding from her.[2] After a few days, the hide can generally be removed. When I noticed that quite a few of the *sarlag* cows were nursing more than one calf, Amaraa explained to me that *sarlag* are good mothers and that they are "kind animals," as they readily adopt calves that are not their own.

Two cows, Khar (Black) and Sergelin (Smart), were mother and daughter. Both cows held relatively high positions in the cow social hierarchy; both would stand near the gate to the pen containing the calves and occupied the same basic position next to each other during milking times:

> Sergelin lost her calf and for the first couple of days afterwards she was clearly distressed, attempting to follow me around and grunting at me, even while hobbled. She is particularly friendly and allows me to scratch through her thick forelock. Chandaga (Arctic Hare) feeds from her mother (Khar) but has also been encouraged to feed from Sergelin with the hide of Sergelin's dead calf on her back. Sergelin has now accepted this surrogate calf and grunts when the calf comes running towards her, even though Chandaga is a different colour from her deceased calf [Chandaga is white whereas the calf was chocolate brown-black]. Tsetsegmaa tried to encourage Chandaga to feed from a third surrogate mother but, in this instance, the calf refused. (Field note excerpt: Choijo's summer encampment, August 17, 2005)

It is important for the herders to have a good knowledge of the heritage of the herd and the social relationships between various members of the herd. Sergelin may have readily accepted having this calf suckle from her, as Chandaga is essentially her half-sibling (they have the same mother, Khar). The calf would have had a familiar smell, similar to that of her own calf, especially with the initial introduction of the dead calf's hide placed on Chandaga's back. An alternative strategy to using the hide of a dead calf is to rub the calf in salt or in the cow's own milk. Herders are a crucial component in this mother–offspring bond, as both mother and young are reliant on the

[2] See Vocalisations, available at www.cambridge.org/9781107000902.

herders for the continuation of their bond through the feeding and milking process.

FOOD INCENTIVES

A mineral that has contributed an unlikely role in the coexistence of herder and herd animal is salt. According to Harris (1996), reindeer, cattle, sheep, and horses all crave salt (Western farmers also provide salt licks for livestock). He suggests that the provision of salt may have been the principle means by which people first tamed and established a degree of control of social herd animals.

An early explorer, Douglas Carruthers, wrote about the Todzha (or Todja) from Uriankhai, in Southern Siberia (present-day Republic of Tuva). He noted that their tame reindeer were free to forage but would return to the tents at dusk for a handful of salt, or to lick the salt from peoples' urine left in the snow. He wrote, "It is a common sight at evening to see the women feeding their pet deer with salt out of little leather bags, as they come home to the tents for the night" (Carruthers 1913; 231). Each reindeer was linked to one person and milked by hand. Similarly, in a somewhat dated paper on reindeer domestication, Mirov (1945) refers to the reindeer-herding Todja as having the "highest degree of domestication" because in his view they were the first to tame reindeer by feeding them salt and milking them twice a day. He suggests that the Todja adopted these techniques from the horse and cattle herders of northwestern Mongolia.

Simoons (1968) describes how the mithans (a bovine) in the Assam hills of India can be persuaded to return from the forests and remain near villages by exploiting their craving for salt. According to Tibetan folk tales, the yak is also thought to have been domesticated through taming by the giving of salt. Tibetan herders still catch yak by offering them salt (Osmaston et al. 1994). Similarly, herders in Mongolia allow their herds to forage for themselves but encourage them to stay near the encampment during the winter months with salt as a food incentive.

Herding families offer food at key times in an animal's life, providing a small amount of supplementary feed when an animal gives birth to young, when an animal is particularly weak in early spring, or to encourage the herd to stay near the encampment during the colder months. Ochero fed a small amount of oats to the Black Stallion every morning at the winter encampment, so the stallion was calm around humans and regularly visited the encampment (accompanied by the rest of the herd). In Choijo's encampment they kept the horse herd nearby by occasionally offering salt, so during winter and spring the herd would visit regularly of their own accord. Food is thus a key element in motivating animals to remain within the domestic sphere, even if the animals are capable of foraging on their own for much of the year.

HABITUATION TO HANDLING

Co-domestic animals accept humans as part of their social sphere through early bonding with humans at formative stages in their development. When discussing how sheep and goats were initially domesticated, Harris puts forward the proposition that domestication "may well have been initiated by the capture and rearing of lambs and kids which, through regular feeding and familiarization with their keepers, would quickly have become tame and would have continued to associate with their human providers when mature, even if they were unconfined" (Harris 1996: 453). From the example of the unconfined animal husbandry in Mongolia, it seems that this is very likely to have been the case.

Associations with humans are one of the animal's first experiences in life. The members of the herding family try to be present at the birth of each herd animal to ensure that the mother has no complications at birth and to remove mother and young under shelter to provide protection from the elements, or potential wolf predation. During the first few weeks the young are handled regularly to assist them in feeding effectively from their mothers.

Herd animals are considerably calmer and easier to manage if handling is a routine, everyday occurrence.[3] At the Arkhangai field site, the horses were habituated to the milking procedure and routine; the mares in Choijo's encampment are milked throughout the summer and into autumn (for six to seven months of the year). At the Bulgan field site, all mares in the surrounding valley are milked for approximately one month in autumn, because the herding families milk only in order to distil a small amount of Mongolian spirit (*nermel*) for the Mongolian New Year, or *Tsagaan Sar* celebration. The contrast between the tameness and ease of control of the mares and foals as a result of the different degrees of handling was clearly evident between the two extended families. The mares at Dogsomjav's encampment were far more wary during milking than the mares at Choijo's encampment, and the foals were considerably more difficult to control. The text that follows is an

[3] See Saikhanaa and the Horses, available at www.cambridge.org/9781107000902.

account of the initial tethering and mare milking of Dogsomjav and Myagmajav's combined herds:

> Boloro milks while the men, Ochero and neighbouring herder Nasaa, handle the feisty foals. Today was only the second day this season that the mares have been milked. The men catch the foals with an *uurga* by containing the herd in a purpose-built corral (unlike foals in Arkhangai, which are tame enough to be caught by hand). The foals are still very wary of being tied to the tethering lines and two of them pull on their halters frantically, attempting to run away but end up flipping over onto the ground. One was so exhausted from its escape efforts that it lay on the ground, heaving for breath. Another was persistently pulling, glistening and dripping with sweat through its fresh winter coat. A mare kept moving away when Nasaa tried to get the foal to nurse from her. Both mare and foal kicked out at him, the mare striking him hard on the inner leg, so that he could hardly walk for a couple of days. (Field note excerpt: Ochero's autumn encampment, October 18, 2005)

After this incident the mare who kicked Nasaa was hobbled by three legs during the day so that she could be approached for milking. Boloro informed me that it generally takes approximately one week for the foals to get used to being tethered and the mares to being milked. Initially, the foals are not used to being led with a halter, so they refuse to budge, or try to escape by bucking and charging forward.

From the ease of handling foals at Choijo's encampment, in comparison to the feistiness of those at Dogsomjav's encampment, it is evident that if young horses are not handled extensively during the milking season, at a formative stage in their lives, then they are considerably more difficult to handle when they are older.[4] Mongolians have a saying: "Tie the horse down as a foal, and train as a colt" (Bayarsaikhan 2006: 221). If the horse is handled early and tied to a tethering line, then it will be easier to train as a two-year-old horse for riding purposes (see the Training section in later text).

A herd animal's behaviour towards the herding family may change within its lifetime, depending upon the amount of handling and daily contact that occurs between human and ungulate. One example of this was with the cow Uuganaa (Firstborn). When I initially visited Dogsomjav's encampment in early March 2005, Uuganaa was the only calf that had been born at that time for that year. She was tied to a post just outside the *ger*, when we visited. Because she was close to the domestic sphere early in life, she knew her

[4] Dogsomjav and Myagmajav's encampments combined ten mares for milking purposes. In contrast, Choijo's encampment milked a much larger group of thirty mares.

name well and was unafraid of humans. Later, in the summer, during milking times, she was required to wait until last to nurse from her mother, as she was larger than the other calves and knew the routine. Because she knew she would have to wait, instead of waiting anxiously at the gate to be released as other calves did, she just sat in the pen, slowly and lazily getting to her feet when finally called to the gate (although, during thirteen milking sessions, Uuganaa always responded to her name by coming to the gate of the pen). As a calf she was *nomkhon* (tame or quiet) but *zalkhuu* (lazy), probably because she was so habituated to human contact.

When I visited in 2007, Uuganaa was three but she had already given birth to a calf, so she was now a young mother. The herding family explained to me that young mothers do not know how to feed their newborn babies, because they are young and inexperienced. Uuganaa had accepted her calf but Dogsomjav told me that her behaviour had changed. "Uuganaa is *muukhai* (bad), she's being *dogshin* (difficult)." Uuganaa tried to butt Dogsomjav as she tied hobbles around her front legs and kicked her udder as Dogsomjav tried to milk her teats. Dogsomjav explained that this change in Uuganaa's behaviour was because she was not yet used to being milked, since it was her first calf. She had only given birth two days prior to my arrival, so Dogsomjav said the milking may be tickling her, or causing her discomfort. To train Uuganaa, Dogsomjav only milked briefly, just enough to get the milk flowing from the udder. At this early stage in the milking season, Dogsomjav milked the cows briefly in the evening but only released the calves to feed from their mothers in the mornings without milking. It is necessary for the herders to not only engage with the herd animals from birth but to maintain that contact throughout the animals' life, training them through important stages in order for the animals to remain calm and manageable in the presence of humans.

To remain "tame" (*nomkhon*) requires constant social interaction with humans and is a crucial element in the co-domestic relationship between the herd animals and herders. In some respects the herd animals in Mongolia are semi-feral, as they are not contained within fencing, ranging freely while grazing, and are able to live according to their inherent social structure. Herders can approach most of the herd animals while out grazing and come within a few metres, especially while on horseback. Herders may be dominant in a behavioural sense but the herd animals have an undeniable trust for the herder, through early habituation to their presence and frequent interaction on a daily basis, especially through daily milking. The herd animals' early engagement with humans means that the herders are essentially accepted as *elders*. Juliet Clutton-Brock states that domestic animals engage in a mutual culture with humans, which is "a way of life imposed over successive generations on a society of humans or animals by its elders. Where society includes both humans and animals, then humans act as the elders" (Juliet Clutton-Brock, personal communication, October 3, 2008).

PART II. CONTROL

> Cultural differences in approach to the human–horse relationship have been evident from ancient and classical history. These differences persist to the present day. There are two main approaches, a co-operative approach based upon understanding the behaviour of the horse, and an alternative approach based on human dominance and submission. (van Dierendonck & Goodwin 2005: 65)

Mongolia is a "land without fences."[5] However, one of the key factors frequently mentioned in the definitions of the domestication process is that the animal is kept in captivity. A commonly cited definition of domestication is Juliet Clutton-Brock's: "A domesticated animal is one that has been bred in captivity for purposes of economic profit to a human community that maintains complete mastery over its breeding, organization or territory, and food supply" (Clutton-Brock 1987: 21). Obviously, this definition does not apply to human interactions with ungulates in Mongolia. Herd animals in Mongolia are not bred in captivity, nor are they kept primarily for economic profit, and herders do not have exclusive control over their breeding, territory, or food supply. Although this definition certainly does apply to factory-farmed animals in Western countries, it does not describe the co-domestic relationship that occurs in Mongolia.[6]

In Bulliet's (2005) "post-domestic" countries, captivity is a requisite for domesticated animals, not because of a threat of predation but mainly because of a lack of space and a perceived need to manipulate an animal's breeding. Goodwin (2002) outlines the costs of domestication, such as restriction of movement, lack of social interaction, and limits to maternal behaviour, yet these costs do not exist to the same extent in Mongolia. Holding herd animals in captivity is not a necessity, nor is it a key factor in the co-domestic relationship between herder and herd animal. The herd animals in Mongolia are remarkably calm and unafraid when engaging

[5] The cover of the Lonely Planet Mongolia guidebook advertises Mongolia as "The Land without Fences" (Mayhew 2001).

[6] Rabey also asserts that this kind of definition does not apply to llama populations in the Andes. He states that llamas "are not removed from their natural habitat, neither are they maintained under controlled breeding conditions … even less is their gene pool under total control of humans" (Rabey 1989: 274).

with humans, yet they are free-ranging animals nonetheless. Manipulation in terms of behavioural modification is kept to a minimum, with herders taking advantage of the animals' normal behavioural repertoire and predispositions.

CONTAINMENT

Because Mongolian herd animals are not fenced in, the herd animals are free to engage socially. This lack of containment allows social hierarchies, bonds, and learning to form naturally without limitations from lack of space or overcrowding. One instance during my fieldwork exemplified to me the herding view on keeping animals in captivity.

> Dogsomjav's grandchildren found a duckling near the calves' pen and brought it back to my *ger* to show it to me, knowing that I would be interested. Dogsomjav suggested that I should release it down by the river, as the parents may be nearby. I carried the duckling down to the river in my pocket to release it but was unable to find any adult ducks. The duckling must have imprinted upon me because it determinedly followed behind me whenever I placed it on the ground and attempted to walk away. A crow was on the opposite side of the river. I thought it would surely be eaten if I abandoned it there. I took the duckling back and placed it inside a cardboard box, gave it grass to lie on, water, and the grandchildren and I found crickets for it to eat. Dogsomjav didn't like the idea of me containing the duckling within the enclosed space of a cardboard box but left me to care for it. The duckling died, probably from being fed too many crickets. (Field note excerpt: Dogsomjav's summer encampment, June 12, 2005)

Dogsomjav's view of how the duckling should be treated differed from my own, as she preferred to leave the duckling to its own devices. At the time I did not consider the box as placing the duckling in a captive environment. I perceived the box as a means of protection, similar to a nest. The duckling did not survive and could have been just as likely to survive if it had been left to its own devices. Here I was viewing the situation as "human knows best," whereas Dogsomjav thought that it was up to the duckling to find its parents, or in other words "ducks know best."

In outlining this example, I do not want to convey the perception that Mongolians do not contain their herd animals at all; rather, when the animals are contained, the restriction is minimal and primarily for protection from predation by wolves, or rustlers, and in response to the weather during winter and early spring. Newborn animals are brought into the family *ger*, and then, when they are stronger, lambs and kids are housed under shelter.

Calves and foals are also housed under shelter in late winter, and early spring, to provide protection in the first vulnerable days after birth. The herd animals, however, are not held captive in an environment that limits behavioural interaction, or that manipulates the animals' circadian rhythms and routines for the purposes of reproduction.

DOMINANCE AND HIERARCHY

Vicky Hearne (2000) and Donna Haraway (2003) conceptualise our relationship with companion species from an animal training perspective, in which it is necessary to be dominant over the companion in order to have a respectful relationship. As Haraway states, "inter-subjectivity does not mean 'equality', a literally deadly game in dogland; but it does mean paying attention to the conjoined dance of face-to-face significant otherness" (Haraway 2003: 41). It is my perception that Mongolian herders are required to retain a leadership role, or behavioural dominance, because otherwise the herd animals would not cooperate at all. Some of the male herd animals, particularly horses and cattle, are considerably larger and stronger than any member of the herding family and would ordinarily be dominant as a result of their size. Handling and control of the male horses or cattle is usually the role of the stronger young men in a herding encampment. When an animal becomes difficult to handle, it is expected that the healthy, younger herding men (roughly between the ages of twenty and forty years) will take measures to control the animal. This involves a cooperative effort, in which all available men assist in holding the animal for tasks such as capturing, hobbling, or training.

Interestingly, even small children are able to assert their dominance with a tame (or *nomkhon*) ox, stallion, or bull. Children have the advantage of being able to use tools, and are able to throw objects, or hit animals with a stick, to make up for their small stature. Children living in the herding encampments attempt to control the animals by holding them down by the horns, picking newborn animals up, or throwing objects, playing at directing the herd. They also have a size advantage once they are up in a saddle on a horse's back. By school age, at the age of seven, they are expected to begin assisting with the daily chores with the animals. The job of taking calves in and out of the pen to nurse from cows during milking is the task of school-age children. During summer it is mainly the children's responsibility to make sure the calves, lambs, and kids do not stray too far from the encampment and that they do not attempt to rejoin the main herds. In this way, both children and young animals grow up socialising with one another and children learn to direct the herd animals, as leaders of the herd, from a young age.

When a herder rides a gelding, the social structure of a horse herd changes. Both rider and gelding are able to take a dominant role over the stallion, and therefore the herd as a whole, because gelding and rider are larger in size and have the aid of tools, namely a whip (*shelbuur*) or lasso-pole (*uurga*). Remarkably, the stallion perceives a difference in status between the same gelding without a rider and in the separate context when the gelding has a rider. One herder on horseback can effectively lead and guide the stallion and his associated mares and young. If there is more than one stallion (which is the case in Choijo's encampment), it requires more than one herder on horseback; otherwise, the horses tend to break off into separate bands in divergent directions instead of remaining as one conglomerate herding unit. Herd social hierarchy is remarkably complex and dependent upon context. If the hierarchy were fixed and linear, then it would be impossible for the herder(s) to break into this social hierarchy as lead individual(s).

TRAINING

Ox Carts

Herders start training oxen (castrated males) to pull handmade, wooden carts when an ox is three years old. A trained ox is generally kept for pulling loads for five to six years. Those castrated male cattle that are chosen to be trained are ideally quiet, light-hooved, quick-stepping (*khungan kholtoi*), and "soft-nosed" animals (*zöölön khamatai*), or easy to pull along (Ulaanaa was one such ox). When the ox is three years old, the herder makes a hole through the nose, or a halter, for catching the ox. Initially, the ox is tied to another cart by a rope through the nostrils but will not pull any weight of its own, merely following along behind another ox. According to the herder, Sanjaa, after approximately ten days of training, the ox can usually carry the weight of a cart as part of a train of oxen.

Choijo's encampment's large herd included fourteen oxen, mostly *sarlag–Mongol* cattle hybrids (*khainag*). A few were well trained and calm around humans, yet others were *emneg*, trained but not particularly easy to handle at all. Nose rings are an effective means to make the oxen more responsive to handling. These fourteen oxen would have ropes threaded through their nose rings when pulling carts from one encampment to another. Any pulling on the nostrils causes the oxen to react quickly, as the nostrils are sensitive to the pull of a rope.

Dogsomjav's encampment had two Mongolian oxen that were well trained and frequently used for collecting water or firewood. These Mongolian oxen did not require the use of a nose ring and were easily caught. A well-trained ox, like a well-trained horse, is of great value to a herding family because he can carry household possessions, firewood, or building materials and can be used to fetch water or ice from the nearby stream for the needs of the household.

Breaking in Horses

A horse may be trained (or taught, *mor' surgakh*) for different purposes: as a racehorse, a riding horse, or even for the capture of other horses (an *uurgach*). Horses are chosen on the basis of such characteristics as bone structure, muscle distribution, their speed as they walk, and their general temperament, particularly those that are "quiet" (*nomkhon*) and "soft-mouthed" animals (*zöölön amtai*). According to the herder Sanjaa, who was particularly adept at breaking in horses, it takes two to three days to train a horse to be ridden. Teenage boys break in the young horses, because the boys are lighter in weight than men. To Mongolian herders the breaking-in process is a test of bravery and mettle for both horse and rider. As Bayarsaikhan describes, "the task of catching and training an untamed horse is a big challenge of might and ability for a young horse herder" (2006: 189).

A horse is initially ridden while still small, as a two-year-old (*daaga*). The technique of breaking in horses was similar in both of my field areas, although it differed slightly depending upon the age or behaviour of the individual horse. Previously, I wondered whether the term "breaking in" would be applicable to the technique used in Mongolia, but I have concluded that it is a suitable phrase, because the purpose is essentially to tire the horse enough that it is unable to fight back against its rider as a result of exhaustion. The Mongolian technique of breaking horses is remarkably successful; After the initial breaking (*unaj*), the horse is responsive to the rider and is able to be directed in a relatively calm manner. The short time span for training is probably due to the horse's intensive handling while being tied to a tethering line as a foal.

> The two-year-old horse was caught with the lasso-pole just as easily as any experienced horse and Ochero put the bridle on without resistance from the colt. The young horse did not know how to move forward when led, so the teenage boy [from a neighbouring encampment] led it like a calf, holding the head and twisting the tail to make the colt move forward. They only tied a single tie around the colt's girdle, which is all that the rider has to hold on to. Ochero held the colt, while the lad jumped on and then said for Ochero to "let go" (*tav*). Ochero slapped the released colt on the rump to make it move forward, while the boy held onto the tail with one hand and the girth with

> the other. The colt moved away rapidly, bucking initially, but quickly calmed down. Ochero followed on a mare [the colt's mother] and both rode as fast as the colt could go, up the valley. Once they returned and the colt was exhausted, the boy groomed him all over with a scraper and then tied him to a post to get the colt used to being handled and "hardened" by standing at the hitching post. (Field note excerpt: Dogsomjav's summer encampment, June 14, 2005)

The *daaga* was easily broken in and quickly controlled, bucking only briefly. The young neighbouring herder explained to me that he breaks the horse in by facing uphill, so the two-year-old tires easily. After the initial bout of bucking the rider trains the two-year-old to respond to a pull on the left side, which means to move away from the stimulus to the right, and vice versa on the opposite side; this is reinforced by the rider's leaning forward and slapping the horse on the cheek, to make it move away in the required direction. Herders refer to this as the horse "learning its mouth."

In Choijo's encampment, some horses are not ridden until they are three, four, or even five years of age because there are too many horses to break in during each horse's second summer. Usually horses are only ridden for the first time during the summer, because they are not strong enough at other times of year. Horses that are broken in at a later age are more difficult to control because they have become larger and stronger than a more pliable two-year-old. Several men cooperate together to capture a "*khangal*," or uncontrollable horse. One person holds the horse still with a lasso-pole; another holds the head still by grasping both ears. Holding the ears and tying a rope around the loose part of a horse's muzzle are also techniques for control in Western riding and racing circles. According to Hart (1985), this is not painful for the animal but works as a type of anaesthetic, calming the horse. Once the herder sits on the horse, the men let go and the rider whips the horse on the hindquarters to encourage it to buck or gallop as much as possible. An audience may gather to watch the spectacle.

Before the horse is ridden, the herders make predictions as to the possible behaviour of each horse, once it is released with a rider. A horse that is difficult to catch may bolt into an uncontrollable gallop when ridden. Preparations include decisions about whether to ride with or without a saddle. A saddle is generally used when the horse is older, as the horse is likely to be difficult to handle while being ridden. The following social dialogue between two herders holding an unbroken horse took place before the young herder, Saikhanaa, mounted and trained the previously uncontrollable horse.

Sumya (to Saikhanaa):	Saw the reins and pull, particularly on the left-hand side.
Buyaa (to Saikhanaa):	If you can make this horse quiet, I'll give you a bottle of vodka and a packet of cigarettes.
Sumya (to Buyaa):	He doesn't need a bottle of vodka.
Buyaa (to Saikhanaa):	Well you could gamble with it, or give it away.
Another herder:	These horses with a thick coat buck well.
Buyaa:	Maybe, but this one just bolts around fast. Be careful. Try to fall off on boggy ground.
Another herder:	I think Saikhanaa won't fall off because he's learnt to stay on now.
Sumya:	I think this one won't buck – it looks so quiet.

(Conversation from digital video tape, Choijo's autumn encampment, September 6, 2005)

In this instance, the horse bolted and did not buck. The group of men stood to watch briefly but left Saikhanaa, moving on to other tasks, as the horse galloped frantically around the encampment. I witnessed Saikhanaa riding and breaking in a number of horses. This was the only occasion where I saw Saikhanaa having difficulty with a horse, as this *khangal* horse would not stop, even when Saikhanaa sawed on its mouth with all his strength.

Wild animals may be perceived to be *out of control* whereas domesticated animals could be perceived as being *controlled* and subjugated. In an ethnographic account of rodeo in the United States, Lawrence states that the horse encompasses the polarities of wild and tame. The process of "breaking a bronc" symbolises "conquest itself, the subduing of the wilderness, the transforming of nature to culture through the process of taming that which was wild and that which was free, as it was enacted upon the American frontier" (Lawrence 1990: 223). Breaking in horses in Mongolia could be interpreted in a similar manner, with an emphasis on a distinct duality between wild and tame, but a Mongolian horse's behavioural temperament does not change with the act of being ridden from that of a wild animal to a tame one. Through the act of being ridden the horse adjusts its position within the social hierarchy to allow the herder to take a lead, or dominant role (if it had not already done so earlier in life). Patton argues that "what we learn from the discipline of animal training is that hierarchical forms of society between unequals are by no means incompatible with ethical relations and obligations toward other beings" (Patton 2003: 95). Essentially, although the herder is exhibiting dominance over the horse that is being ridden, this does not mean that the rider is not treating the horse with respect. In fact, the horse demands respect because it is a strong, powerful animal that cannot be subdued easily. The Mongolian herder works within the horse's own social boundaries and disposition to become an accepted part of the horse's social hierarchy. According to Tumerjav, in Mongolia "there is a tradition to train livestock, particularly horses, without brute force, but rather with gentle positive actions. This is an important trait of Mongolian animal husbandry" (2003: 105).

PUNISHMENT

Western training techniques use a basis of punishment and reward. In Mongolia, herders rarely punish the herd animals, and they do not reward them with food items. Interestingly, Anderson writes of a similar attitude amongst Saami reindeer herders: "The reindeer herding dog, like the Saami child, is treated roughly but affectionately, and given little negative feedback. Instead being rewarded for unique, independent and innovative behavior, both dog and child ripen into maturity at their own pace, picking up skills as needed or desired" (1986: 10). Mongolian children, like the herd animals, are expected to gradually learn from trial and error and experience, rather than coaching and instruction. Herders successfully control the herd animal's behaviour by giving vocal signals, accompanied by body language (for details, see Chapter 5), rather than forcing the animal to comply.

In Choijo's encampment the members of the extended family were more inclined to use objects to exert control, hitting a goat on the horns with a stick for example, thumping a cow with a milking stool, or striking a horse with the lead rein if they were being difficult during capture. This hitting was not in anger, as the herder hitting or kicking the animal would not behave in an uncontrolled manner but was clearly asserting himself as the dominant individual and reinforcing the offending animal's submissive status. This was not a general method used, however, as I did not see the herding family in Dogsomjav's encampment using objects, such as a piece of wood, to maintain dominance. One of the men in Dogsomjav's encampment occasionally hit the animals with force with his *shelbuur* but Dogsomjav expressed her opinion to him that it was unnecessary to hit the animals in such a manner. Perhaps one of the reasons for this difference in the methods of control is that Choijo's encampment has much larger herds, meaning it is harder to control all animals to the same extent. Dogsomjav's encampment has a herd of considerably smaller size, making it easier to handle the herd as a whole. An analogy could be drawn with a teacher and a class full of school children. As any teacher could attest, a class of fifteen children is easier to maintain control over than a class of forty, or more. Teachers can also vary in their attitudes and have different techniques for gaining control over a class. In my view, the Mongolian herder takes a leadership role analogous to that of a teacher, whereby the pupils are not necessarily subjugated or their freedom undermined but there is a societal expectation to abide by the direction and instruction from a more knowledgeable role model, or elder.

CONCLUSION

Herd animals in Mongolia socialise within a herd and are not bound to respond only to humans; they have autonomy and are not expected to perform as slaves or automatons. Mongolian herd animals are initially part of the household when young and remain amenable to social interaction with humans, but they are not bound to the domestic sphere because they are released to roam around and graze during the day. The herd animals may be tame and are controlled to a certain extent, but they are not coerced against their will, nor are they unable to carry out a social existence with other members of their own species. Mongolian herders successfully take the lead role as a member of the herd, and they effectively take advantage of the mother–offspring relationship to obtain milk from female animals. However, this is done with the motivation and active cooperation of the herd animals themselves.

PART THREE

LIVING WITH HERDS

7

IN THE LAND OF THE HORSE

> The centrality of the horse to human existence on the central Eurasian steppe throughout the Holocene is reflected by its virtual omnipresence in the cultural and economic life of steppe peoples. (Levine 2004: 115)

> For the Mongols, a man without a horse was a nobody; a man was virtually never seen in the countryside without his horse. Furthermore, a man was known by his horse. Mongols did not judge a man by his clothes or his accent. They looked carefully at his horse, and by its proportions, color, gait, and by the look in its eye they could tell all the essentials about its owner. Knowing this, people chose and educated their riding horses with almost unbelievable care. (Waddington-Humphrey 1974: 485).

INTRODUCTION

Mongolia is described in popular literature as "The Land of the Horse," conjuring up an image of a nomadic herder galloping across the vast grassland steppe. The Hollywood ideal of the cowboy could easily be conceived of as descending from the myth of the nomad (Khazanov 1994). When in Mongolia, the dust-swept towns and the rugged-looking men riding in on horseback brings to mind the archetypal western scene. In this chapter I explore some of the ways in which the horse is of central cultural importance to Mongolian herders. The complex role of the horse is a fascinating aspect of Mongolian herding society, especially because the horse is both a utilitarian resource, as Mongolian herders eat horsemeat in winter, and a unifying symbol, expressed on Mongolian flags. Unlike the other parts of this book, Chapter 7 focuses solely upon people and their horses and does not encompass other human–animal relationships within the Mongolian co-domestic sphere.

Throughout history, the horse has occupied multiple roles and purposes in human society. Horses were a key component of early warfare; in many areas they were used as draught animals to plough fields, and as transportation, either ridden or used to pull carriages (Clutton-Brock 1992). Eurasian horses' military role made them "the jeeps and Sherman tanks of ancient warfare" (Diamond 1997: 91). Gradually the horse's role in Western society has changed: The automobile has replaced the horse as a primary means of transportation; the tractor as a means of traction; and the tank in warfare. In many Western societies, the horse is no longer eaten as food and is mainly engaged in leisure activities, such as horse events or general "hacking."

Animals still occupy multiple and complex roles in other "domestic" (or co-domestic) cultures (Bulliet 2005). One animal takes precedence over other alternative domestic animals, such as the yak in Tibet (Downs & Ekvall 1965; Ekvall 1961, 1968; Miller 1999); cattle in African nomadic pastoral societies (Olupona 1993), such as the Nuer (Evans-Pritchard 1940, 1956); reindeer in Siberian ethnic groups (Ingold 1980, 1986b, 1994; King 2003); or the camel for nomadic desert peoples, such as the Bedouin (Gauthier-Pilters & Dagg 1981). A feature of these key domesticates is that they aid in people's mobility, exhibited competitively through racing. Camels are raced in North Africa and in India, whereas reindeer are raced in Siberia. These societies show other linking characteristics, in that the predominant animal occupies multiple roles. These animals are valued both in a practical sense but also spiritually; there is intensive contact with the animals on many societal levels; and all live in extreme environmental conditions, so consequently human groups are reliant on these key animals for subsistence and even survival. This pattern of a key domesticate also characterises the horse in Mongolian herding society. This human–other animal relationship is built upon layers of meaning and tradition, resulting in a complex interactive mutuality between two species: human and horse.

When I arrived in the Mongolian Khangai Mountains in March 2005, it was the start of spring. I expected to interact with horses on a daily basis and imagined that I would spend all day out herding on horseback. I was disappointed when I realised that the horse herd was grazing far from the encampment, and only one horse was tied to the hitching post, available for herding the sheep and goats. I had a lot of contact with the cows and calves during milking, and with the sheep and goats and their newborn lambs and kids. I expected herders to express how important horses were to them but instead they would talk about the importance of the other herd animals. For example, Lhagva said "we exchange clothes or flour for our sheep; goats are important because we need the cashmere and it is profitable in Mongolia . . . but I like cows the most because I milk them." She did not mention horses at all.[1]

[1] Horses, and particularly the competitive activity of horseracing, also tend to be a male-oriented sphere of interest in Mongolia (refer to Chapter 8).

This contradicted my preconceptions, as popular travel literature on Mongolia emphasised the historical importance of the horse and its prominence in Mongolian art, music, and folk legends (Altangerel 1987, 2006; Becker 1992; Stewart 2001).

By the beginning of June the routine of herding life changed. The mares had given birth to the first foals of the season. In Arkhangai, herders began milking the mares to produce a fermented milk product (*airag*). Contact with the horse herd became far more frequent: The foals were attached to tethering lines, and the mares required milking every two or three hours throughout the day. This was also the time when the training for the summer festival (*Naadam*) season began, approximately one month before the first race. It is at this time of year that the horse visibly becomes an integral part of Mongolian social and leisure activities. The mares play an important role in providing milk for an alcoholic beverage that provides necessary vitamins that cannot be obtained through other sources (see Chapter 8). Geldings are ridden to herd encampments for social visits, further facilitating social interactions between neighbouring herders.

In the *Naadam* the horse is given immense status and privilege over other herd animals, linked with the status of the herding male, a historical link with warfare, and mobility as a crucial aspect of Mongolian nomadism. Although my initial experience of the level of importance placed upon the horse did not meet my expectations, the horse is indeed held in high esteem. This is best revealed during the warmer summer months when life is less oriented towards subsistence and more towards social engagement and prosperity. I aim to demonstrate that the horse holds both a symbolic and a physical presence in Mongolian society, both as a being in the world and as a behavioural ideal, exhibiting mobility, strength, and stamina. I aim to demonstrate that the horse acts as a "key symbol" in Mongolian culture (Ortner 1973). The following section gives an indication of the long association of human and horse in Inner Asia. Even early remains indicate that the horse had significance in death, as a means of carrying a person into the afterlife.

PART I. THE HORSE IN MONGOLIA'S PAST

> It is not possible to imagine Mongolian history without horses, and I think it is not possible to view the future of Mongolia without horses as well. Mongolia is not Mongolia without horses. (J. Tserendeleg, president of the Mongolian Association for Conservation of Nature and the Environment)[2]

[2] The quote is from the American Museum of Natural History Web site: http://biobulletin.amnh.org/D/2/2/index.html (retrieved September 4, 2004).

> In Mongolia, herding is nomadic, so the horse is very important. Herders need horses. Historically, horses were important during the time of Chinggis Khaan for military purposes. Horses were like tanks. If they [a people] had good horses then they were a strong force throughout the world. Horses are very famous for having a better mind than other animals: they can recognise their own children well and separate the herd, with young going to another stallion. All people in Mongolia like horses! During summer time I enjoy riding horses. *Hiimori zurikh*, it supports my mind [or, it lifts my spirits]. (Divangar, Ulaanbaatar, November 3, 2005; spoken in English)

As I cantered across an expansive river terrace on my way to visit a herding encampment in a neighbouring valley, I came across two large stone blocks with inscriptions of leaping figures carved onto the surface. These stone blocks, facing east, are referred to as deer stones (*bugan chuluu*). It is estimated that there are approximately 550 of these stones scattered across the grassland steppe of northern and western Mongolia. Most deer stones date back to the Bronze Age. The images depicted on these stones are ethnographically similar to the symbolism of shamanistic practices of the reindeer peoples of Siberia. Deer stones predominantly depict leaping deer but also feature horses. These stones jutting out of the surrounding grassland reminded me that this land had been inhabited by herders riding around on horseback for a very long time.

Archaeological research indicates that a nomadic horse culture lived near the Altai mountain range on the western edge of present-day Mongolia over 2,400 years ago. At a site in Pazyryk, archaeologists unearthed immaculately preserved tombs, kept frozen by permafrost throughout the centuries. The continuity between these herders and the herders of Central Asia today is remarkable. The people subsisted much as Mongolian herders do today, relying on horses, cattle (including yak), sheep, and goats. This was evident from the types of clothing they were wearing, and from felt products and small food offerings found within the burial chambers (or barrows). The people lived in felt tents or log cabins and had shelters for their young animals in the colder months (Rudenko 1970). Among the items of felt is the oldest known carpet, with illustrations of a man riding a slender horse. The horse was particularly significant in their daily existence, in both transport and traction but also spiritually, for carrying the deceased into the afterlife (Samashev & Francfort 2002).

The Russian scientist Sergei Rudenko excavated a number of burial mounds in expeditions in 1923 and again between 1947 and 1949. These large mounds (or *kurgans*) were the burial chambers of important chiefs, or elite clan members, and included the embalmed bodies of either men or women with elaborate clothing and intricate tattoos. Sacrificed horses had been stuffed with dry grass and embalmed to preserve the bodies. As many as seven to sixteen horses were sacrificed in each burial chamber; all were castrated males, or geldings. Rudenko (1970) reasoned that mares and stallions were likely to be too important to sacrifice. As geldings would have been used for riding, they would have been more appropriate to carry the dead into the afterlife. Each sacrificed horse wore an elaborate saddle, bridle, and head regalia embossed with gold. These headdresses were adorned with reindeer antlers, mountain goats, and griffin figurines. Remarkably, the Pazyryk bridles have the same basic design as the bridles used throughout Inner Asia today (see Clutton-Brock 1992, Figures 7.4 and 7.8).

It is evident that the people of this horse culture had a preference for specific horse colours, used marking techniques for identification, and clipped horses' manes in a manner similar to that of Mongolian herders today. Horses within one burial chamber also had a variety of earmarks, similar to the techniques still used to identify cattle, sheep, and goats. Each gelding's mane was clipped short and the tail was tied in a manner similar to that used for Mongolian racehorses of today in preparation for a festival. Even the coat colour of the horses was preserved within these frozen burial mounds, revealing that the horses were mainly chestnut, brown, and less commonly bay, with a distinct absence of white leg markings. One *kurgan* contained horses that were all dark duns, another with light sorrel coats (Nickel 1977). The size and stature of the horses suggests that they are related to the Mongolian horse breed that lives in the Altai Mountains at present (Samashev & Francfort 2002). There is debate on the origins of the humans in the burials. Initially it was thought that they were closely aligned with, and perhaps related to, nomadic Scythians who also sacrificed their horses, but many had features that are similar to Mongolians today (Azzaroli 1985).

Burial mounds of a different cultural tradition also exist in Mongolia. Turkic *khirigsuur*, or burial stones, are accompanied by circular or square burial mounds. These too provide an indication of the significance of the horse, because they also contain sacrificial horses for carrying the dead into the afterlife. According to ethnographic accounts from the neighbouring province of Tuva, the custom in which a horse was killed and buried with its rider continued right up until the nineteenth century (Vainshtein 1980).[3]

[3] Horses are no longer sacrificed in Mongolia, because of the influence of Buddhist practices that prohibit animal sacrifice, as well as the secular influence of communism.

THE MONGOLIAN HORSE AS CONQUERER

The horse gave Mongolians as a people the mobility to expand out from their homeland, travelling over large distances to invade the territories of distant kingdoms but also with the capacity to return home. Much of the migration of the peoples of Central Asia was likely to have been from the ease of travel by horseback along an east–west axis, rather than north to south where there are greater geographical barriers for horses, such as deserts and mountains (Diamond & Bellwood 2003). The speed and endurance of the horse meant that Mongolian armies could achieve far greater feats than would be possible on foot.

The Chinese were the first to compose a written record of a nomadic people that occupied present-day Mongolia and North-western China, the Xiongnu (or *Khunnu,* meaning "First People" in Mongolian). As protection against the Xiongnu, the Chinese established an early version of the Great Wall, not to keep out the "barbarian" nomads but to halt their horses' progress. The European name for these nomads was the "Huns" (similar to the Mongolian "*Khunnu*"), and most notable was the leader Attila (who ruled during the fifth century). In Europe, Attila was referred to as "the Scourge of God" by those he conquered (Baabar 2004). The Xiongnu paved the way for the largest empire in history, the Mongol Empire.

In the year 1206, the leader Temüjin, who had united warring nomadic tribes, became known as Chinggis Khaan; the name meant "'Almighty Ruler'" (Onon 2001). This was essentially the birth of the Mongol state (Khazanov, 1994). Chinggis Khaan (1155–1227) successfully united neighbouring warring factions to form an unstoppable force, invading and conquering the territory of China but eventually capturing a region extending from present-day Hungary to Korea and from Siberia to Tibet (Roux 2003). Ultimately his descendents ruled the largest empire in history. By 1241 the Mongol army had conquered on horseback as far as the edges of the Volga River, but upon hearing that Ögöödei Khaan (son and successor of Chinggis) had died, the men in the army turned their horses around and returned to their homeland (Baabar 2004). Ironically, their strong sense of family ties and connection with Mongolia as home meant that the Mongols did not conquer as far as Western Europe.

Soldiers were able to survive on horseback for months by tapping blood from the neck of their mounts or drinking fermented mare's milk, and they did not require supplementary food for their horses, because the Mongolian horse could scrape through snow to eat the vegetation beneath (Gladitz 1997). Chinggis Khaan established the Great Law (*Ikh Zasag*), documented in *The Secret History of the Mongols* (Onon 2001), which outlined the customary laws of the day. Soldiers were prohibited from hitting their horses about the head, and the stealing of horses was punishable by death. The Mongols established an efficient communication network across the vast empire through a horse relay system (*örtöö*), a precursor to the Pony Express mail system in the Midwest of the United States (Roux 2003). This successful mail system lasted for over 700 years, from the thirteenth century to the mid-twentieth century (Tumerjav 2003).

The Mongolian Hordes' success depended on specific technologies of the day: the horse, stirrups, and the composite bow. However, because of the mobile nature of the Mongolians and their horses, few physical traces of the Khaan dynasty are left upon the Eurasian landscape today. The Mongol people and the horses who accompanied them left their heritage in the genes of the populations they conquered, but few archaeological remains are left throughout what was once a vast empire (Baabar 2004). Nevertheless, many cultural traces remain: in language, in literature, and in the mythic imagination of the West. Technological contributions include the process of making felt (Laufer 1930) and the use of bacterial cultures to make different kinds of dairy products (Indra 2003). The nomadic Mongolians and their sturdy horses are likely to have had a greater influence upon Eurasian culture than is generally acknowledged.

Specific types of horses were held in particular esteem by the Khaans, or noble elite, and were given the title of the "top herd." They were selected on the basis of power, quality, and aptitude (Tumerjav 2003). This herd would include a selection of the best ambler horses (*joroo mor'*) for riding purposes, chosen for their soft and gentle parallel gait (Erdenekhuyag 2004; refer to Chapter 4). It is evident that preference for specific horse colours was important in the time of the Khaans. In war, horses were grouped according to coat colour, as they were representative of the cardinal directions. The five cardinal directions of Chinese cosmology (and the corresponding horse colours) are red for South (red sorrels); black for North (black); blue for East (essentially grey – the closest approximation to blue in horse colours); white for West (white); and yellow for the centre (golden–palomino). As The Great Khaan was the centre of the cosmological universe, he had horses of a colour appropriate to his status. I saw many family *ger* with carpets that depicted the eight palomino horses of Chinggis Khaan. These eight horses have great spiritual and symbolic significance in their role of protecting The Great Khaan (Erdenekhuyag 2004).

The Selenge River is important historically because it is the largest river in Mongolia. Coincidentally, the Selenge River is within a few kilometres of Dogsomjav's winter encampment at the Bulgan field site. Beneath a prominent rock outcrop on the edge of the Selenge, local herders get off their horses to prostrate before the rock face. The local veterinarian (Aagii's father) told me this rock outcrop is sacred because it was a resting and watering place for Chinggis Khaan's army and their horses. The figure of Chinggis Khaan and his warfare on horseback looms large in the history of Mongolia and the world.

HORSE CULTURE

The horse is present in all facets of human life: in birth, in death, in work, and in leisure, as nourishment and as spirit being. In Mongolia the horse represents wealth and status, inspiration, identity, mobility, and continuity with a long and prosperous historic past. The idea that A. D. King relates from the Kamchatka region is true for Mongolia as well. King (2003) writes how a reindeer herder explained to him the importance of reindeer: "Without deer there is no culture, nothing." In Mongolia the power and presence of the horse is equally central. In herding encampments, children grow up riding horses from a young age. They begin herding the sheep and goat herd on foot but soon graduate to herding on horseback. Herders spend long periods of time communicating only with other animals, particularly the horse they are riding while out herding. The nomadic herding lifestyle in Mongolia could be an isolated one, with such a large land area with so few human inhabitants, but with the ability to ride on horseback to neighbouring encampments the herders have mobility and are able to sustain a close herder social network. The horse basically provides the link between the close-knit extended herding family, or home, and the outside world.

A wonderful example of the great power and spiritual significance of the horse is the analogy made between the horse and the Mongolian musical instrument the *morin khuur*, or horse-head fiddle. The horse's head on the fiddle is regarded as a talisman, according to Mongolian spiritual belief, so great care is taken in carving the wooden head of the horse. The instrument itself is imbued with power because it has the ability to create emotion through the music it produces. Historically, the base of the instrument was made from horsehide and the handle of the bow from the rib of a horse. The two strings and bow are still made from horsehair. The larger "male string" is made from 130 hairs of a stallion's tail, whereas the smaller "female string" is made from 105 hairs of a mare's tail. The tail hair is to be taken from a horse ridden since childhood, and each hair fiber should be processed until it "starts talking."[4] The sound of the *morin khuur* can be reminiscent of a horse galloping across the grassland steppe, or a horse's whinny. When learning

[4] This is from the Mongolian Art & Culture Web site: http://www.mongolart.mn/traditional_instruments.html (retrieved September, 2007).

the *morin khuur*, students learn how to play the gaits of the horse from walk through to gallop. Compositions on the *morin khuur* praise the horse's gallop, the colours of the horse, its temperament, sharp intellect, courage, and loyalty.

I was told a number of versions of the legend for the origin of the *morin khuur*. It is clear in this legend that the horse is a pivotal link for returning home from afar and a means of bringing people together, in this case two lovers. The following is an extract from one version of the legend that was recorded during Sven Hedin's expedition to Mongolia in 1943. The hero of the story would visit a herding woman every night on a mystical horse with wings, then before dawn he would ascend into the night sky as a morning star. The herding woman clipped the wings of the horse to try to keep her lover by her side but to no avail:

> After the flying prince had been rushing through space on his marvellous horse for a time, on the way to his star, he discovered that his steed was beginning to show signs of exhaustion. And finally, in a great desolate desert, the horse sank to the ground and died. In despair over the loss of his marvellous steed, and persuaded that he would now never again be able to reach either his star or his beloved, both of whom were so far away that he could not reach them on an ordinary horse, he began to caress his dead steed. And between his stroking hands, while the tears fell from his eyes, a miracle took place. The dead horse was transformed into the first morienhur [sic], an instrument that was ornamented on top with the horse's head; and the horse's mane and the hairs of his tail, which he had been holding in his crushed hands, were changed into the sounding strings.... From that time forth, the prince wandered over the steppes playing his *morin khuur* and singing about his distant star, his lost love and his precious horse. (Jenkins 1960: 130)

In war the horse's speed allowed for a capacity to conquer and control that was unique in history. At the same time, it is these same qualities that enable people to come home. For a nomadic people these are qualities that hold together all the desires for closeness, for love, for overcoming separation, and for mediating between worlds.

The horse, for Mongolian herders, nourishes the body and the mind, expressed in the arts and in the herder's imagination. The traditional shamanic symbol of the horse is depicted with flames above the horse's back and beneath the hooves. This symbol is still featured on Mongolian flags denoting health, goodness, and wealth (Damdinjaviin 2005). The ritual of a horse carrying a person into the afterlife has been adopted into the practice of Lamaist funerals. Davaa informed me that when her father died, the previous year, a lama designated that only light-coloured horses should be ridden in the funeral procession to the grassland gravesite near their summer encampment. This is also reminiscent of the selection of horses of a particular colour in the Pazyryk burials already discussed. As early as 2,400 years ago, the ancient burials in the Altai Mountains revealed that sacrificed horses were a part of peoples' funeral practices (Rudenko 1970). Although horses are no longer sacrificed alongside herders when they die, the care in selection of horses for funeral processions indicates the continuity of the horse-flight-afterlife matrix: Horses carry a herder's body to the burial site, symbolically carrying the person into the afterlife.

The horse has clearly been important in the lives of herders for thousands of years. Mobility and dispersal in conquest takes people to other places on earth, and death takes people to another place that is not on earth so flight becomes necessary. In both cases the horse provides a means of mobility and thus a means of moving from one place, or world, to another. The horse has enabled Mongolians to become who they are today; without the horse, they would not be Mongolians. Many of the ancient traditions just described resonate in the ritual and ceremony surrounding horseracing festivals in Mongolia.

PART II. THE *NAADAM*

In this section, I examine the horse's importance in the competitive event of horseracing because this is a distinguishing element that separates the horse's role from other co-domestic ungulates in Mongolia. The *Naadam*, meaning "festival," has an ancient heritage stretching back to the warring time of Chinggis Khaan. A meeting of the different "banners" every second year, or *Ikh Khural*, meant that wars could be organised and politics discussed but this also meant that soldiers could demonstrate their skills and prowess through the "three manly sports" of wrestling, archery, and horseracing. Banners made from the tails of 1,000 white stallions were used to signify peace, whereas the tails of 1,000 black stallions were used to signify war (Erdenekhuyag 2004). These banners are still traditionally used within the Mongolian parliament today and are displayed every year in the opening ceremony of the national *Naadam*.

The festival season is a time for local herding families to celebrate the warmer, prosperous season that the summer brings. The temperature is above freezing, so the grass is experiencing new growth from the summer rains, and horses are energetic from the new grass in their bellies. The first type of festival is the *Övöö* festival,[5] held in celebration of a time when the

[5] An *Övöö* festival is held near a prominent mountain. Local monks make blessings near a prominent rock cairn (*övöö*) and herders gather to make offerings. This festival is held

rain falls and the grass begins to grow (Tomikawa 2006). Visually, a *Naadam* is full of colour and motion in the midst of a stark, seemingly limitless landscape. Horses are attired with silver embossed saddles and bridles, while the trainers themselves wear trilby hats perched at jaunty angles and their best clothing (*deel*).[6] Red and blue flags flutter in the wind, people are dressed in brightly coloured silk clothing, jockeys dress in dazzling capes and pointed caps, and the horses' coats shine. Most spectators arrive at the *Naadam* just before the first group of racehorses, the stallions (*azarga*), cross the finish line. The herders do not tend to station themselves at various points along the race but wait at the finish line, resting on the backs of their horses. Small children are propped up in the saddle in front of a parent, boys double bareback, and unmarried young women wear their best *deel* and ride beautiful horses. Elderly herders dismount from their horses to sit on the grass and discuss the state of grazing and recent weather conditions while exchanging small bottles of snuff (*khamrin tamkhi*) and bowls of fermented mare's milk. People and horses gather together in an energetic throng of noise and colour, contrasting vividly with ordinarily life in the herding encampment, which is quiet and leisurely (see summer *Naadam* photographs, Plates 11–14).

The crowd whoops and yells as the first five horses cross the finish line; then spectators and trainers rush to where the horses are already circling to cool down. This is unlike the distinct segregation in English horseracing between participants and spectators with a defined class structure (Cassidy 2002). Mongolian spectators are members of an encampment and one of the racehorses is likely to be from their own herd. This results in a very different type of involvement from English racing circles, in that without class distinctions the spatial line between participants in the race and spectators merge at the end of each race.

The mobility made possible by the horse is represented in each race, where horses and riders depart from the same point to which they return (unlike Western horseracing, which is on a circular track). It takes approximately one hour to ride the racehorses to the turning point and then to turn and race back to the finish line, which means that for spectators there is approximately a two-hour interim between each race. While the crowd waits for the next race, everyone disperses and boisterously gallops on horseback the kilometre or so to the wrestling stadium.

Wrestling is a competition to exhibit both strength and endurance, which is also the function of horseracing. Wrestling bouts are not time limited and do not depend upon weight. The wrestler, analogous to the racehorse, may

by local herders and is not as large as the town or provincial *Naadam*, which are officially held to celebrate the Mongolian revolution of 1921.

6 As nomads, who carry all their possessions with them to new pastures, Mongolian herders traditionally displayed their wealth as silver decoration on their saddles and bridles.

have a body representing a beast or a bird of prey, containing supernatural strength. Tomikawa (2006) states that the Mongolian wrestler is a symbol of the community; likewise the racehorse epitomises the strength of the herd. The spectacle of wrestling and horseracing is indicative of the herder's ideal of strength, power, fearlessness, and physical ability exhibited by the Mongolian male and the Mongolian horse.

In Bulgan *Süm*, instead of sitting on the rickety wooden seating around the edges, most people sit watching the wrestling from horseback, elevated above the stadium fencing. Riders lean on the backs of neighbouring horses, and horses rest their heads on the rumps of other members of their herd. The rider occasionally signals for the horse to move forward or back, occasionally engaging in conversation with others while watching the wrestling spectacle over the fence. The interconnectivity between human and horse is an important part of the sociality of the *Naadam*. Formal performances are accompanied by intimately textured, unplanned physical interconnectivities between horses and humans.

SELECTION

A horse is selected for racing based on unusual coat colour, or specific physical and behavioural features. What follows here are three examples of ideal criteria for selecting a good racehorse:

> The best over-five-year-old horse should have a rectangular head, big eyeballs, long neck, rectangular body, curved back, the chest and hindquarters must be angular, thighs must be thick, the crutch wide. It must have nine whorls on the front section of the belly and eight whorls on the back section.
>
> A fast three-year-old foal should have a wide body, four thick legs, straight back and stomach, wide hooves, and plentiful eyebrow lashes, feel heavy when trotting and have angular teeth. It is a good sign if it lies down when the sun shines on its four legs, when it smells its own dung, looks back at people when they pass and when it mates with mares, all these characteristics are good.
>
> The best foal, as a baby, should have the body of a jerboa,[7] a calf-like muzzle, bulging eyes, ears like a goat, and a raised tail like a calfs while gamboling, with four prancing legs. If it has all these characteristics, then it will be a good horse.[8]

[7] A jerboa is a small desert rodent with very long hind legs and a long tufted tail. A foal is described as analogous to a jerboa (*alagdaakhai*) if it is particularly fast. The foal should have long hind legs as a physical indication of its potential.

[8] I was unable to find the author of this text, but it was a slender leaflet used by Dogsomjav's deceased husband to select racehorses and contained horse *tamga* on one side (see Chapter 4, Figure 4.4), with the horse characteristics on the other. Similar leaflets could

From these defining characteristics, herders are obviously very perceptive about a horse's physical form: Even the direction and number of whorls on a horse's coat is important. The references to other animals indicate that favourable characteristics are appropriated from other species to form the ideal horse. It is also significant that selection is based on a variety of factors, including physical characteristics, how the horse lies, where the horse is placed within the herd, and how the horse responds to human presence. The sign of a good racehorse (*khurdan mor'*) is even imprinted in the tracks it leaves upon the ground, for instance:

> When examining the horse by making it gallop a short distance, the marks of the horse should be even in one line. Generally, the horse that leaves marks of the front edge of the hooves and does not leave marks of the heel is said to be a fast horse. If the marks of the four hooves are distinct, and the hind [hoof marks] are left above the front, it is considered to be a very bad horse. (Bayarsaikhan: 2006: 232)

The Astrological Wheel

Types of Mongolian racehorse have been further categorised according to an astrological wheel configuration, and that can be illustrated here through a table with a grid-like structure (see Table 7.1). This information is based on seventeenth-century books (or *sudar*) that discuss the characteristics of the different racehorses (Telenged 1996). The astrological arrangement can be divided into three broad types of horse, related to the sky or heavens (*tengeriin khüleg*); the wind (*calkhin khüleg*); and the earth (*shoroon khüleg*). Characteristics of the sky-type horse, for instance, are as follows: it is like the sun and the moon; it leaps like a lion through snow; it has a fine chest; long front legs; an uplifted head (*tolgoi deegüür*); a curved muzzle; a sharp-edged backbone; and it is a soft-mouthed animal. The ancestry of Mongolian horses is then further divided up according to three origins: (1) those aligned with a feathered bird (*jigüürten yazguurt khüleg*), with features mixed between a large bird and a butterfly; (2) those aligned with an antelope (*göröösön yazguurt khüleg*); and (3) those aligned with a beast of prey (*araatan yazguurt khüleg*). A horse originating from the feathered bird-type of racehorse, for instance, is recognised as being somewhat timid in crowded situations; with an "honest" (*shudarga*), upright gait while walking; when ridden it follows under guidance but will fly once released; and, when tired, this horse will gain energy back rapidly. Each category of horse is associated, on

be bought in small shops near Buddhist monasteries for herders to purchase but did not tend to credit an author.

Table 7.1. Astrological table for racehorses

Origins	Dwelling in sky or heaven (*Tengeriin ornii*)		Dwelling in space (*Zavsriin ornii*)	Dwelling in the land (*Gazriin ornii*)	
	Feathered-bird origins (*Jiguurten yazguurt*)		Antelope origins (*Göröösön yazguurt*)	Beast-of-prey origins (*Araatan yazguurt*)	
Dwelling in the atmosphere (*Ogtorguin ornii*)	Griffin-like sign–nature (*Garid shinjit*)	Dragon-like sign–nature (*Luu shinjit*) Camel sign–nature (*Temeen shinjit*)	Gazelle sign–nature (*Zeeren shinjit*) Wild ass sign–nature (*Khulan shinjit*)	Mountain sheep sign–nature (*Argalan shinjit*) Deer sign–nature (*Bugan shinjit*)	Lion sign–nature (*Arslan shinjit*)
Dwelling in the wind (*Salkhin ornii*)	Demoiselle crane sign–nature (*Togoruun shinjit*) Bird sign–nature (*Shuvuun shinjit*)	Snake sign–nature (*Moroi shinjit*)	Billy goat sign–nature (*Ukhnan shinjit*) Musk deer sign–nature (*Khuder shinjit*)	Jerboa sign–nature (*Alagdaakhai shinjit*) Racing dog sign–nature (*Taigan shinjit*)	Fox sign–nature (*Ünegen shinjit*) Rabbit sign–nature (*Tuulai shinjit*)
Dwelling in the earth (*Shoroon ornii*)		Fish sign–nature (*Zagasan shinjit*) Frog sign–nature (*Melkhii shinjit*)	Wild horse sign–nature (*Takhi shinjit*) Saiga antelope sign–nature (*Bökhön shinjit*)	Elk sign–nature (*Khandrai shinjit*) Cattle sign–nature (*Ükher shinjit*)	Wolf sign–nature (*Chonon shinjit*)

Note: The table indicates the grid-like structure for the ancestral origins of the different types of "natures" of racehorses.
Source: Adapted from a diagram in Damdinjaviin (2005).

a third scale, with twelve different animals, both real and mythical, based upon their physical and behavioural features (note that the years, months, and days of the Mongolian calendar are similarly divided up into twelve different animal categories in a parallel fashion; see Chapter 10). Those racehorses dwelling in the sky or heavens (*Tengeriin ornii khüleg) and* originating from a feathered bird (*Jigüürten yazguurt khüleg*), for instance, can be further categorised, according to appearance and behaviour, to a demoiselle crane (*togoruu*); a griffin-like creature (*garid*)[9]; a dragon (*luun*); or a frog (*melkhii*). In contrast, those that dwell in the earth (*Gazriin ornii khüleg) and* originate from a beast of prey (*Araatan qazguurt khüleg*) are categorised as resembling a lion (*arslan*); a racing dog (*taigan*); a rabbit (*tuulai*); or an elk (*khandgai*). By studying such configurations, a horse trainer can look at a herd and, according to a particular horse's behaviour and physical characteristics, he can establish that horse's specific ancestry. From this categorisation of the horse's origins the trainer can, therefore, assess the horse's potential as a racehorse, or as a stallion for breeding future racehorses.

These ideal characteristics, however, were not as immediately evident as criteria for the herders that I questioned. One of Ankhaa's racehorses that had just been broken in was a chestnut two-year-old with a blaze. When I asked Choijo why this particular horse was chosen to race, he said it was because he was "*aashtai,*" or moody, and the intensive training would make him calmer. Choijo pointed to another chestnut that was being trained for the over-five-year-old (*ikh naas*) race and commented that his behaviour was similar when he was a young two-year-old. When Ankhaa groomed the sweat from the two-year-old after training, the feisty young horse kicked out with a hind leg, or shoved Ankhaa with his head. The two-year-old was indeed *aashtai.*

TRAINING AND RACE PREPARATION

Horseracing in Mongolia is a race of endurance. Mongolian horses are not swift sprinters but are hardy and have the stamina to cover large distances. These characteristics of stamina and endurance are important for surviving the extreme winter temperatures in Mongolia; for use of the horse as a means of mobility over large distances; and in the historic role of the horse in warfare (Baabar 2004). Incidentally, one relay in the horse-based mail system, which still existed in Mongolia until as late as 1954, was the same distance

[9] This griffin-like creature (*garid*, or *khangarid*) is also featured by the praise-singer when he is referring to the winning racehorses during the "Praise Ceremony" in the section below (also see Mayor, 1991).

as the current length of the national championship race, a distance of thirty kilometres (Tumerjav 2003).

To increase a horse's stamina, the horse trainer monitors the horse's intake of food and water. He ties the horse to training posts every day, for one month in the heat of summer, before the first race of the *Naadam* season.[10] This act of hardening the horse for training has existed at least since the sixteenth century. According to tradition (as depicted by Surruys 1945: 149, cited in Bold 2001), the reasoning behind this training regime is to "solidify the fat" on the horse:

> After 3 or 5 days, or after 8 or 9 days, the fat on the horse's back has thickened, its belly is small but strong, its croup is large and solid. Before it was a false fat which [resulted] from the green grass; now the fat has grown solid and thick. [The horse] can gallop at full strength without losing its breath. The horse can participate in battles, can go without sufficient water or grass for 8 or 9 days without exhausting its strength.

These methods are used to "harden" the horse, as herders believe that mental preparation and the loss of weight increases a horse's endurance and stamina. Horses are generally tied to a hitching post (*mornii uyaa*) before and after riding, but separate posts are erected specifically for training racehorses (*khurdnii mor' uyakh*), and these posts are established approximately forty metres from the *ger*. This separates the everyday utilitarian use of the herding horse from the festive activity of horseracing. Training posts are symbolically significant, as they connect the horse and rider to the encampment (the co-domestic sphere) as the point from which both always depart and to which they both return.

Before any training exercise the horse and rider circle around the training posts at least three times in a clockwise direction. During this process, the rider sings "*Giingoolokh*," which is thought by herders to act as mental preparation for both horse and rider. Training around the posts must be done in the correct manner and with respect for the success of the *Naadam* season. No person is to pass under the line that runs between the posts; instead, humans and horses are expected to walk in a clockwise direction around to the opposite side (even though it would be easier to merely pass beneath the tethering line).

"Sweating the horse" is a crucial part of the training process, used to increase the horse's strength and stamina for racing.[11] Each horse is trained

[10] The length of training is variable, depending upon the health and age of the horse but also the preference of each trainer (see Table 7.2 for variable training lengths depending upon the age of the horse).

[11] See Training for *Naadam*, available at www.cambridge.org/9781107000902.

Table 7.2. Race categories: Distances of racing groups of horses

Racing group		No. of days required to train for racing	Distance of race (km)		
			Long-distance practice	Summer *Naadam*	Total (start–finish)
Stallion	*Azarga*	25	15	22	44
5+ years	*Ikh Naas*	25	20	25	50
5 years	*Soyoolin*	20	12	22	44
4 years	*Khyazaalin*	20	8	18	36
3 years	*Shudlin*	15	6	15	30
2 years	*Daaga*	15	5	10	20

Source: Some information is from Bayarsaikhan (2006).

with quick galloping sprints (*tarlalt*) of one to three kilometres, intended to increase the horse's lung capacity so that it may endure rides of longer distances. Adult trainers (*uyaach*) sometimes ride the horses during training, especially when sweating the horses (*mor' khölsökh*), but young boys are the main riders. "Closed sweating" involves covering the horse in a layer of felt while training. In the heat of summer this causes the horse to sweat profusely, increasing the horse's metabolism and causing it to lose weight for racing. After a sweating session, the trainer may taste the horse's sweat to interpret how to proceed with further training. "Muddy sweat" tastes salty, whereas "watery sweat" has no taste at all (Bayarsaikhan 2006). After each training session the horse is cooled down by circling around the training posts, and the trainer removes the sweat with an elaborately carved scraper (*khusuur*). The act of scraping off the sweat functions to open the pores in the skin, removing any waste from the horse. This training routine is essentially similar to the actual act of racing, so both horse and rider become physically and mentally prepared.

Training is the domain of men and boys and provides a chance for friendly rivalry between neighbouring encampments. Boys are usually between five and twelve years old, which means they are not too heavy for the horse to race, yet skilled enough to push the horse to its limits. Children are encouraged to learn to break in young horses alongside racing the horses. A young jockey will usually train and race bareback, wearing socks without shoes. It is said to be harder to race with stirrups, and racing without is thought to reduce the chance of injury to both the horse and child (Bayarsaikhan 2006). Preparation extends to practice races (*sungaa*) between neighbouring encampments, giving the trainers a chance to gauge the horses' progress and to give "green" horses and jockeys a taste of competition. These practice gatherings become social occasions within themselves. When I asked Saikhanaa to summarise how a horse is trained for a *Naadam*, he replied, "at first they [the trainers] choose the horse, then tie it to the hitching post. A few days later I try galloping. Yesterday, after trotting a number of times around the hitching post, I tried racing (*tarlalt*). Tomorrow we will have a practice race near where the festival will be held" (Saikhanaa, summer encampment, August 18, 2005).

On the actual day of a *Naadam* the horses are prepared for racing in a specific manner that is not applied to horses in everyday circumstances. Before travelling to a festival, and during races, the racehorses have their forelocks and tails tied up. Ganbar, as a trainer, was very particular about the way the leather tie was to be wrapped around the horse's tail and asked his father, Choijo, to retie it when it was not tied exactly two-finger-widths apart. I was given a number of reasons as to why trainers tie up the racehorses' forelocks and tails: for decorative purposes, "because it looks impressive"; to ensure that the horse's vision is not obscured (particularly in windy conditions); and to prevent the tail from becoming entangled in the horse's legs, decreasing the horse's speed. I suspect that it is also representative of the horse's status as a racehorse (*khurdnii mor'*).

Early deer stones from the Bronze Age depict horses with objects protruding from the head in a similar manner to the tufted forelocks of racehorses (see Sorokin 1990). Sacrificed horses from the Pazyryk burials had elaborate head dresses and their tails were tied up with cloth (Rudenko 1970), which was similar to the triangular pieces of cloth tied onto the tails of Mongolian racehorses. Contemporary ornamentation of the horse with head regalia appears to signify speed and flight, similar to the representation of horses in sacrificial burials, representing passage into the afterlife, established as a symbol thousands of years ago.

In preparation for a race, the trainer carefully grooms each horse with his elaborately carved scraper (*khusuur*) and curry comb (*sam*), which a herder wears to show that he is a trainer. The *khusuur* is an object with power through its crucial link between trainer and horse and its use in the scraping of sweat from the horse. The scraper has images carved on the surface depicting auspicious Buddhist symbols. The *khusuur* was traditionally made from the beak of a pelican (a coveted material, as Mongolia is so far from any ocean) but is now made from wood. On race day, the trainer attaches the scraper and comb around his waist by using a blue silk scarf (*khadag*) that signifies the power of the scraper. Herders sometimes carry a simple *khusuur* slipped down the inside of their boot and use it while out herding when the horse they are riding is moulting, or if the saddle has rubbed, but this is not attached to a *khadag*. Mongolian horses are not generally groomed for aesthetic reasons and are only pampered during festival occasions.

In Choijo's encampment, before a *Naadam,* Ganbar blesses each horse with incense, burning a locally wild, aromatic juniper plant (*arts*) in an incense burner. The burner is circled around the neck three times and then held to the horse's nostrils so that the horse breathes in the aroma. Before departing for a festival, incense is burnt near the training posts and the racehorses are prepared for departure by conducting the ritual of circling around the training posts in a clockwise direction. The racehorses are then led by the trainer, to conserve their energy, to the location of the *Naadam* several kilometres away.

THE RACE

There are six race categories during a *Naadam,* each based on the sex or age of the horses that are competing. Before each race a horse's age may be checked by examining the teeth. Stallions (*azarga*) usually race first and young two-year-olds (*daaga*) tend to race last (refer to Table 7.2 for race categories). The order is from stronger, older horses to younger, less experienced horses. The age of the rider is not significant because it is the horse that is valued for its speed and endurance, not the skill of the rider. If a horse without a rider gallops through the finishing flags within the first five, the horse is still counted as a winner. The stallion race holds a high level of status for a herder, as the stallion is the signature and pedigree of the horse herd as a whole. It is amazing to see young children of four or five years riding stallions that are not just symbols of strength and power but are physically much stronger than a human, particularly a small child. While racing, the stallions look impressive as they gallop across the expansive landscape with long, flowing manes. *Daaga* are still not fully grown and may still suckle from their mothers before or after their race. They are usually led, or merely follow behind their mother, to and from the *Naadam,* while the trainer rides the mare. The two-year-old that comes last in the field is given a medal to award the young horse for eating well and is praised to bring it success in future races.

Before each race, a herder rides around holding aloft a red flag while he circles around on horseback. The young jockeys gather and follow on their racehorses, singing a loud and elongated "*Giingokh,*" just as they did in training preparation but louder and with more urgency to mentally prepare themselves and the excited horses. This song is only sung to horses and is reminiscent of a Buddhist chant in its repetitive sound and intonation. The herder with the red flag then departs from the point to which they will end the race, guiding the horses and riders at a fast trot across the ground that they will cover during the race.

The ride to the turning point can be tough on the young jockeys because the riders may become bunched together and the horses have to be restrained from breaking into a gallop. The strong and wilful horses may become out of control and start galloping off in the wrong direction, or throw the young jockey onto the ground. Once they have turned to race, the jockeys yell "*goree, goree*," meaning "go, go," to encourage the horses and give them strength. Winning does not just depend upon spurring on from the rider but also relies importantly on the will and tenacity of the horse.

When the horse and jockey cross between the red and blue finishing flags, they are cheered by the crowd and each placing is yelled out by judges from the sideline. The jockey's job done, he immediately dismounts from the horse. The trainer takes the reins from the jockey, grooming the sweat from the horse with the *khusuur*. I would observe the jockeys left standing alone while the horse was groomed and walked around, again in a clockwise direction. The young boys look suddenly small, as if they have been decapitated from the powerful body of the horse that had just been a part of them. Spectators run a hand over the winning horses to absorb some of the horse's sweat for good luck and future prosperity. Both the scraper and the horse's sweat hold a spiritual power beyond a mere object and a bodily fluid; they contain the essence of the power and endurance of the racehorse. Similarly, people wipe some of the sweat off a winning wrestler to gain some of the wrestler's physical fitness and luck.

In Arkhangai, the ride from Choijo's summer encampment to the *Naadam* featuring local herding encampments (*bag*) was approximately six kilometres, whereas the *Naadam* at the administrative centre (*süm*) was eighteen kilometres away. This meant that if a horse and jockey were participating in the over-five-year-old (*ikh naas*) age category, the total distance covered by the horse in one day would be approximately ninety kilometres (refer to Table 7.2 for race distances). As an experienced jockey, Saikhanaa would often ride in three age classes, always without a saddle or stirrups. This means Saikhanaa could have potentially been covering up to as much as 180 kilometres on horseback during a single *Naadam*. That morning he would still be expected to work, hobbling cows and releasing calves during the morning's cow milking. When I asked him whether he felt tired after a full day of racing, he firmly said "no, I'm not tired." To me, the endurance of both racehorses and young riders seems remarkable. Such stamina in both horse and rider could only be attained through daily riding, habituation to hard work, and adaptation to harsh environmental conditions. Swiftness, strength, and endurance are central to the mobility of the Mongolian horse and mobility is central to who Mongolians are as a people, as nomads. The merits of the winning racehorses are exemplified in a praise ceremony that follows each race.

THE PRAISE CEREMONY

A praise ceremony is held after each race to honour the first five winning horses, or the "*airag* five," in reference to the prestige of fermented mare's milk.[12] Each horse is given a name to recognise its status. The winner of the race is given the title of *tümnii ekh*, or "leader of ten thousand."[13] A man recognised for his skill in announcing the horse's virtues will sing for the horses, using a melodic style of poetry. The praising ceremony features an older style of Mongolian and incorporates Tibetan words. The metaphors and symbolism are unique to nomadic herders, although the praise singer may vary his wording slightly during a ceremony, adding his own style to the poetic blessing. The following is an extract of the praise given to winning horses during the Bulgan Süm *Naadam* on July 8, 2005:

> ...Ooh, these horses possess all the features of speed,
> Ooh, these striking horses, with wide eyes, enclosed ears, fine legs,
> Flared, trumpeted nostrils, and black eyes with binocular vision,
> Pearly white teeth with canines like steel, a tall and slender body,
> Elongated 6th vertebrae and withers, wide hips with broad chest and ribcage,
> Ooh, the golden bodies of these horses,
> Have the power of the griffin, with their strong muscles,
> Ooh, these horses leap throughout the world and look afar,
> With eyes as sharp as wolves and an ability to see distant sights,
> Galloping over mountain ranges, across wide valleys,
> With bridle jingling, their ears listen to distant sounds,
> Ooh, with golden reins raised, these horses will always be praised,
> And win every festival race,
> Ooh, the horses bring glory to their trainers throughout the Nation,

12 See *The Naadam* for an example of a praise ceremony, available at www.cambridge.org/9781107000902.

13 Wrestlers are given similar titles after powerful animals from Buddhist texts, such as "Elephant" for the wrestler who wins six rounds, and "Lion" for seven rounds, the winner of the tournament. At a national level, each wrestler has an older assistant who praises his attributes and the number of rounds he has won, similar to the praising of the winning horses. "Leader of ten thousand" is in reference to the horse's ability to lead a large herd but may also be a reference to the horse's role in Mongolia's warring past.

Make jockeys proud and bring joy to the festival audience,
With a clean forehead, silk reins, wings of a swan, wild ass dust,
These young, tough, strong horses are of this prosperous State.[14]

Many of the words and phrases would not hold as much meaning for Mongolians from urban centres, as many of the terms are unique to Mongolian herding culture, especially those relating to the physical characteristics of the horse. The winning horses are compared to admirable characteristics in other animals, such as the strength of a lion or the sight of a wolf. The winning horses are described as having a "clean forehead" because the horses that are further behind become covered with dirt from the dust stirred up by the lead horses' hooves. "Wild ass dust" is a reference to the Asian wild ass (*Equus hemionus*), still existing in the Gobi desert, where herders only catch a glimpse of the herd as they rapidly gallop away and leave only dust in their wake. The griffin (*khangarid*) is a mythical bird-like creature and is recognised for both its strength and power. According to Mayor (1991), the legend of this bird-like creature with four legs originated from the region of Mongolia and spread across to ancient Greece.[15] Speed is metaphorically linked to the concept of flight, with the praise ceremony depicting images of the horses having "wings of a swan" and mythical powers.

In every praise ceremony I witnessed there was historical reference to the winning horses galloping across the world and reaching distant lands, much as the Mongolian horse did in the era of Chinggis Khaan. The praise ceremony successfully conjures up images of the horse as a spirit being with an ability to fly across vast lands, much as Mongolian shamans are capable of flying on the back of a spirit being while in a trance. There are references to valued nomadic possessions, such as gold, steel, and silk. "With golden

[14] In the Soviet era (1921–1990), horses literally belonged to the State. Now, herds have been privatised, so they belong to individual herding families. However, the translation of "prosperous State" should not be taken too literally, as some phrases have been adopted from communist terms and have become integrated into colloquial Mongolian language. My interpretation is that it means that the horse represents the local region: a prosperous environment, producing fine horses.

[15] Mayor (1991) suggests that griffin folklore arose from ancient observations of rich dinosaur fossil beds found along the trade routes linking Central Asia with Europe. She proposes that peoples in the Altai Mountains found fossilised remains of *Protoceratops*, with prominent beaked skulls and four legs, associated with rich gold deposits, leading to the belief that a mythical creature was the guardian of gold. These four-legged griffins have also been found on tattoos of human remains in Pazyryk in the Altai Mountains, alongside gold artifacts featuring beaked quadrupeds (also see Rudenko 1970). Roy Chapman Andrews located rich fossil deposits of *Protoceratops*, along with fossilised dinosaur eggs in the Gobi desert after he heard local Mongolian folklore about dragon bones (Andrews 1932).

reins raised" is a reference to the ceremony itself, because the praise singer holds the reins of all five horses aloft. The praise ceremony is significant in that it directly acknowledges the prestige of the horse during times of prosperity and addresses the winning horses themselves, extolling their physical virtues, rather than their human trainers or the young jockeys.

Fermented mares' milk (*airag*) is an integral part of the praise ceremony. When the poetic praising has finished, a bowl of the white *airag* is passed around. The colour white is symbolically important within Mongolian herding culture, as milk products are essential food sources. The horse has a small amount of *airag* poured onto the crown of its head and rump. Both young jockeys and trainers take a sip from the bowl and the jockeys again sing "*Giingokh*" as they depart. The sprinkling of *airag* upon the horses, and the consumption of it, is an indication of the importance of this alcoholic beverage as an integral part of social occasions in Mongolia. Mares' milk provides nourishment for young horses and people, giving them strength and inextricably linking the two species, binding them together with the sacred beverage.

In local *Naadam*, herders with winning horses are awarded carpets for the interior of their homes (*ger*), whereas young jockeys are given a certificate and soft drink. During the Soviet era, instead of monetary rewards for hard work or elevated status, the State would give medals and awards of merit.[16] This practice still persists in the *Naadam*, where winning horses are awarded with a medal. The fastest horses often win many races. The accumulated medals are displayed on a red triangular-shaped piece of cloth that hangs over the horse's forehead, and this is displayed on the horse after a race, or after an award ceremony. At other times of the year, the medals are displayed, pride-of-place above the shrine, towards the rear of the family home.

The horse–human relationship in Mongolia is all about the extended family encampment. The *Naadam* festival provides the opportunity for herding families to show off their finest horses. The quality of the horse and the skills of the rider are representative of the overall social status of the herding family. According to Tomikawa (2006), the horseracing and wrestling within a *Naadam* is not merely a game, or sport, because the participants are regarded as being possessed with spiritual powers and are to be revered accordingly. These are not just everyday horses that hold the same status as a gelding used for herding, or a mare that produces young. Through exhibiting their superior capabilities in strength, power, and endurance, they are revered, as spirit beings. The praise ceremony is a key means by which

[16] In Mongolia, mothers are awarded levels of merit to encourage women to have a large number of children and ultimately increase Mongolia's small population size (Goldstein & Beall 1994). Dogsomjav has been awarded the highest award for merit, "The Order of Glorious Motherhood, First Grade," because she has ten children.

herders exhibit their reverence of the horse. The herder calling out the poetic praises of the horses extols the virtues of the five winning horses to the audience, pointing out their fine features and superior powers, as spirits of the herd, and therefore the herding family, that they represent. As the praise ceremony expresses, the elite racehorses reflect the power and might of the unbeatable combination of Mongolian herder and horse.

CONCLUSION

The expression of the herder–horse relationship is represented beautifully during the summer *Naadam* festival. In everyday herding practice in Mongolia, herders tend to treat horses in a functional manner, much as they treat the other co-domestic herd animals. Nevertheless, the value Mongolians place on the horse, although formed in part through everyday utilisation and the working roles that horses occupy, is vividly expressed and invigorated during times of ritual celebration. For seven months of the year the temperature does not reach above freezing and life tends towards subsistence and survival for Mongolian herders (Goldstein & Beall 1994). This is contrasted with a short four-month growing season during summer, when herders celebrate their relationship with the horse through horseracing and the milk the horse produces, in the form of the alcoholic beverage, *airag*. The *Naadam* is an event that involves active participation and social engagement on horseback. It is a time for celebrating the mobility, strength, and endurance of both man (through wrestling) and horse (through racing). The *Naadam* reinforces the level of prestige that horses have held throughout Mongolia's long historic past. The horse is an integral, inherent part of Mongolian nomadic culture. I will leave the final words to Enkhtuvshin, a Mongolian friend, who wanted to emphasise to me the importance of the horse to Mongolian people: "An important aspect is that from childhood herders and horses live together, love each other and understand each other. The horse is a clever animal; they understand people well."

8

THE CYCLE OF LIFE: BIRTH TO DEATH, SPRING TO WINTER

INTRODUCTION

In this chapter I describe significant events and activities during the herding year, starting with the birth of the newborn animals in early spring, until the slaughtering of animals in autumn before the onset of winter. In describing spring through to the start of winter, I depict not a full calendar year but at least the main period of herder–herd animal interaction in Mongolia. During the core winter months from mid-November until mid-February, herding families and herd animals resume a period of semi-hibernation in which activity is kept at a minimum to conserve energy so that they might survive the extreme winter conditions. For the rest of the year, daily activity is largely dictated by the routine of the herd animals and by the changing environmental conditions through the change of seasons. In this chapter I provide etho-ethnographic examples of how the herd animals influence particular herders in their day-to-day lives.

Herders are very aware of the changing seasons, weather conditions, and variations in temperature, as their livelihood depends upon it. Each season is very distinct, with dramatic temperature changes and differing weather conditions marking each new season.[1] The average temperature (the midpoint between the daily high and low) in Mongolia is below freezing for seven months of the year (Goldstein & Beall 1994). In winter, the landscape is white with only patches of bare soil on the windswept mountainsides. Every day is short; the sky is blue but the sun barely warms the land. Snow falls mainly in late autumn and in spring, as during the winter the temperatures become so cold that there is not enough moisture in the air for snow to

[1] In Bulgan (near Dogsomjav's encampment), from 2001 to 2008 the average number of days below 0 °C was 242.4, whereas the average number of days below −17° C was 106.7 (www.weatherbase.com).

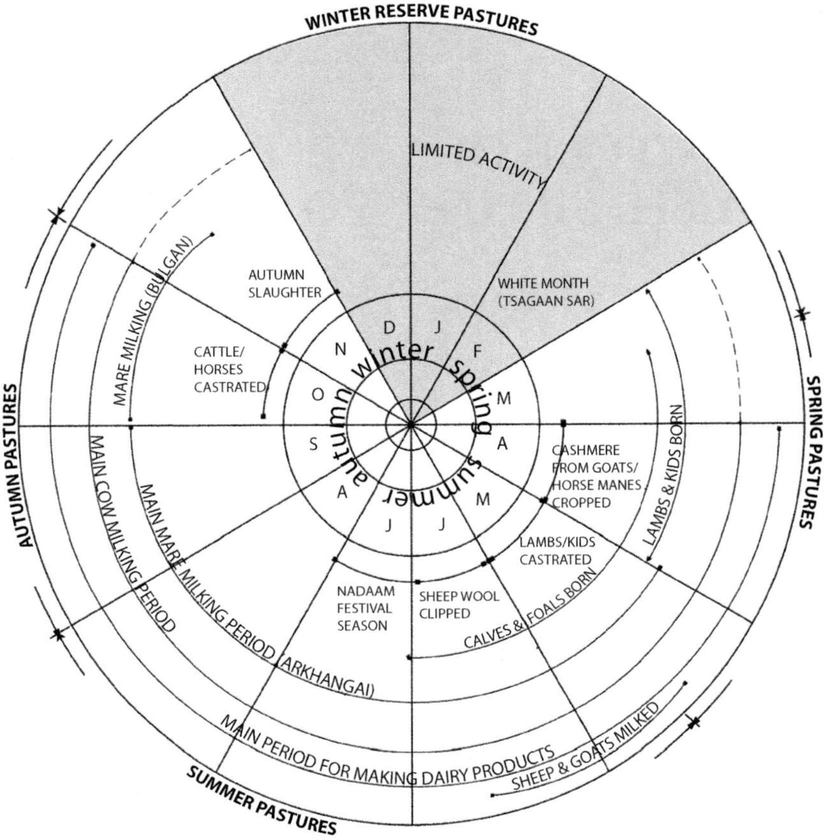

Figure 8.1. Seasonal calendar: Herding life in the Khangai Mountains (events and activities vary, depending on factors such as weather conditions and state of pasture).

fall. Each season is very different from the next, and this means the activities of the herders and herd animals change quite markedly with the four seasons.

Herders are more attuned to a cyclical rather than a linear conception of time. Herders move around their home, a sacred cairn, or a horse hitching post in a circular, clockwise direction. Similar to a *ger*, or a Buddhist astrological calendar, I have depicted the seasonal calendar as a circular diagram, showing the annual cycle of herding activity (see Figure 8.1). For Mongolian herders, the movement to new pastures coincides with a change in the routine of the herd animals, which in turn matches the changing seasonal conditions. As Evans-Pritchard (1939) points out, for the pastoralist Nuer, it is the needs of the cattle that translate what he refers to as the "oecological rhythm" into the social rhythm of the year. This can be illustrated by focusing on two crucial activities in Mongolian nomadic herders' lives: the seasonal movements to new pastures and the milking routine. These forms of migration and milking practices have been recorded in Mongolia since the thirteenth century (Bold 2001). I illustrate how important the milking process is in Mongolian herders' (or milkers') everyday lives, as milking dictates daily routines. I focus on four herders: Choijo, an older herder who is the patriarch of his family encampment; Saikhanaa, a young herding boy, who works for Choijo's extended family; Dogsomjav, an older herder who is the matriarch of her extended family encampment; and Lhagva, a young herding woman, one of Dogsomjav's daughters. These four protagonists provide examples of the different roles herders occupy.

SPRING ACTIVITY

Mongolian herders recognise the coming of the New Year with "White Month" (*Tsagaan Sar*) celebrations, often from mid-February. The date changes from year to year on the basis of the Mongolian calendar (the calendar is described in detail in Chapter 10). Herders say *Tsagaan Sar* (in mid-February) is in celebration of the time when female herd animals begin to produce milk, which essentially presages the coming of spring. The dairy products served during *Tsagaan Sar* celebrations represent purity and goodness to Mongolian herders. This time of year, between the frigid depths of winter and early spring, when the grass has not yet sprouted through the permafrost, is called the "declining period" (Tomikawa 2006). This is a crucial time for livestock because there is little nourishment from pasture and their fat reserves have been gradually depleted during the winter. Actual spring temperatures and weather conditions do not tend to arrive until mid-March. In mid-March there is a dramatic change in the weather; within a matter of days the air temperature suddenly increases by 20° C from the usual winter temperatures below −15° C (Miyazaki, Yasunari, & Adyasuren 1999). The first animals of the year may be born any time after *Tsagaan Sar* until early June, but most lambs and kids are born around mid-March (see the insert of spring photographs, shown in Plates 1–8).

Spring is highly changeable: One day the sky is perfectly blue and the weak heat from the sun begins to melt the frozen ground; the next day there is a snow storm bringing a coating of snow, whitening the barren landscape and endangering the lives of the newborn lambs and kids and their weakened mothers. Spring is punctuated with winds that blow down the valleys from the Siberian Arctic, picking up fine dust particles along the way. Everything becomes covered in dust and ticks, which ride on the winds, lodge in people's hair, and burrow into fleece. Herd animals, already weakened from the sparse vegetation available during the long winter, can die from tick infestation, so herders twist out the ticks between finger and thumb. Even when the snow melts, the ground is still frozen solid. Ground vegetation is reduced to the toughest, standing stalks, which provide meager nutrients for hungry livestock.

It is only in April, when there are the first rains, that new shoots begin to sprout on the higher slopes. While herding, Dogsomjav pointed out two plants that appear earlier than other plants in spring: *Gechgene* (*Potentilla acoulis*), a low-lying carpet plant with small, yellow inflorescences; and *Yargui*, a purple mountain flower, which has a strong smell and is effective against parasites. While herding at this time of year, the goats scatter up the mountain slopes in search of these plants. This makes herding difficult, as the sheep remain grazing together on the lower slopes, while stray goats rapidly disappear over the ridgeline of the nearest mountain in search of the next *Yargui* plant. Herders enjoy eating goat at this time of year, as the meat is flavoured by the fragrant spring plants that they eat.

Saikhanaa emphasised the difficulties of herding in the springtime by saying, "spring is very difficult because when I'm herding the sheep and goats, they can fall through [the ice] into the river and into big crevasses. If they can't get out quickly, then they will die." In spring, many more newborn animals enter the world on a daily basis, requiring care and attention from all members of the herding family. In describing herding sheep and goats in spring, Saikhanaa added, "I always go with them and when some of them give birth, I collect the new lambs and kids and bring them home. After that I go back out with the sheep and goats" (Saikhanaa, summer encampment, August 18, 2005). With the birth of animals in spring comes the production of milk so that female animals can feed their young. Newborn young may die if they are not accepted by their mothers, or if they are too weak to feed themselves. Some do indeed die soon after birth, but most are assisted to feed from their mothers, or are fed by hand until they are strong enough to eat grass (see Chapter 6).

The whole family works hard with the sudden influx of newborn young. Lhagva also emphasised the difficulties of spring to me when she said, "new calves grow into adults with great difficulty because they become overcome by cold easily. We have to look after pregnant cows very carefully" (Lhagva, autumn encampment, October 21, 2007). An example of a birth in spring was Oodon, the Kazakh cow, giving birth to her sixth calf, Menget (Birthmark). The link between human and calf begins from birth. Lhagva assisted Menget by initially removing the birth sac that still covered the calf's head. She then blew into the calf's ears, in an attempt to assist the calf to breathe its first breath in the outside world. The first few days after a female gives birth, she produces colostrum (*uurag*), milk that is particularly thick and yellow. The herding women milk some of this colostrum from the cow and it is eaten by the family, particularly elderly or pregnant women because it is full of nutrients. In turn, the women give the cow some flour and salt, or medicinal herbs mixed in with milk to encourage any placenta that still remains inside to be released and to help the cow to gain her strength back after birth. From the day of birth, the calf encounters humans and learns not to fear members of the herding family. Most of the young calf's contact with humans is with women, as they are the ones who initially nurture the young and handle them during milking times (and, especially if the calf is female, will continue to engage with her daily throughout her life).

Similar to female animals with their young, the herding women are occupied by nurturing their own babies or young children. As Ingold states, aligning with Lhagva's comment earlier, domesticated animals are not "'made'" but "'grown'"; "they are 'mothered', nurtured, assisted – generally cosseted and helped along. Secondly, growing plants and raising animals are not so different, in principle, from bringing up children" (Ingold 2000: 86). In the year I was in the Mongolian countryside, in 2005, Dogsomjav had the addition of four new grandchildren and became the primary carer of the youngest of these babies. She also assisted in nurturing many of the young animals. Caring for young, therefore, takes up a large portion of Dogsomjav's life. In Mongolia, mothers are revered: There are numerous songs about mothers, and male herders are expected to respect and care for their mothers. A baby within a herding encampment, similar to the newborn herd animal, encounters its herd animal counterparts from a very young age. For the first six months of life, a Mongolian baby is swaddled for much of the time inside a blanket and lies on a bed inside the family *ger*, which means the baby's mother can continue to work on chores around the *ger*, knowing the baby is safe. Within this domestic sphere the baby experiences young animals, the newborn lambs, kids, and calves who are also nurtured inside the *ger*. The sounds and smells permeate throughout the *ger*, providing the baby with an immediate associative experience with herd animals.

Babies are highly valued. Caring for and nurturing young (whether herd animals or humans) is an integral part of Mongolian herding society. In Mongolia, the youngest child is the one that remains to care for the parents and the youngest male is given the family *ger*, or "hearth," upon marriage. This attention towards the youngest child is evident when this child is given food before older children. The order in which calves are fed is the same. The first calf is often named Uuganaa (Firstborn), whereas the lastborn is named Otgono. When the women were milking, they would ask me to release the calves from the pen from the individual that was the youngest,

Otgono, through to the oldest calf, Uuganaa. Within a Mongolian herding encampment, both children and young animals begin to learn their place within a relative hierarchy, which is sensitive to age because the youngest are inexperienced and vulnerable.

Haircutting and Combing

In the subsequent text I provide examples of the significance of haircutting as activities for co-domestic people and animals. The cutting of horses' manes and the combing of cashmere from goats is recognised by herders as an activity to be conducted in spring. The Mongolian calendar is consulted for the correct day and time to perform the haircutting activity. Clipping a horse's mane is a job reserved for male herders. Choijo told me that "on the twentieth of April every year, I always cut the horses' manes and tails. At this time, family and neighbours come to help with the haircutting" (Choijo, summer encampment, August 28, 2005). A horse's mane is treated much like a person's hair. Most of the mane is trimmed but a small section is left long around the withers and forelock. Male herders spend a lot of time trimming the horses' manes to look neat and tidy. When cattle (including *sarlag*) or horses are killed, or sold, a lock of their hair (*jakhdag*) is also kept and tied to rope that runs from the centre of the roof of the *ger*. This helps the herder to retain a physical memory of particular individual animals, in a similar manner to the hair from a child's first haircut.

A Mongolian child has his or her first haircut at the age of three to five years, and this act is celebrated in the form of a haircutting (*daikhavakh*) ceremony. The family consults the Mongolian calendar to celebrate the haircutting on an auspicious day. A boy's hair is cut on a separate year from a girl's. This ceremony is an important event in any Mongolian's life and, according to Naraa, is given greater emphasis than a wedding. All the neighbouring herding families and relatives of the child are invited for the occasion and each person cuts a lock of the child's hair. A boy's long hair is tied up in plaits and bows, but after the haircutting ceremony male herders tend to have their hair cropped short (sometimes using the big shearing scissors used for clipping sheep wool!). The haircutting ceremony marks the point at which the child is no longer recognised as a baby but as a boy or girl with a specific gender, indicating the child's entrance into social life. Relatives give the child his or her first animals as gifts on this occasion. In the past, children were given a horse of a designated colour (Waddington-Humphrey 1974: 479, 486). The ceremony is in recognition that the child has survived beyond the vulnerable initial period of infancy in which mortality is high in the Mongolian countryside (Goldstein & Beall 1994). After a child's haircutting ceremony, a lock of hair is kept as a reminder of this significant period in a person's life.

An annual spring activity is to comb the cashmere undercoat from each individual goat. Herding families, including men, women, and children, all help one another with the labour-intensive job of combing each goat. This is a communal activity and each herding encampment staggers their combing with that of their neighbours. When it is time for the members of an encampment to comb their goats, all the neighbouring herders arrive to help in the activity. It takes a number of days to finish combing all the goats within the herd because it may take up to an hour to comb the cashmere from all parts of a goat's body. The goats themselves are not ambivalent about this activity because they only experience this kind of treatment once a year. Some goats bleat loudly with each stroke of the metal comb, whereas others are soothed by the action and relax, closing their eyes. After the combing for the day has finished, one of the women from the encampment will cook a meal to feed everyone in exchange for the work they have put in. A few days later, the encampment will go to the neighbour's encampment to help to comb their goats. Combing the cashmere from goats can be undertaken by men, women, children, and the elderly. Everyone is expected to participate in this social event and to reciprocate the work done by neighbouring herders. This is an example of people joining in regardless of sex or age, because this is a task that everyone can easily participate in.

Tasks are divided by gender or age because certain body types are more capable of performing different tasks. Bold (2001: 65) confirms this when he states "at which age [children] perform tasks, especially boys, depends on the nature of the animals – their age, size and docility – as well as on the degree of difficulty of tending and on the labour potential of the nomadic community." Herders are required to be resourceful and self-sufficient because it is not possible to have a great deal of specialisation; a herder is required to learn herding activities for four (including the yak, five) different species of animal and engage in most tasks, such as training the animals, clipping the wool, or repairing fencing and shelters. The division of labour in daily life complements gender, age, and species-specific domains.

Women tend to interact with female animals through the act of milking, and with young or sick animals that needed nurturing, whereas men tend to interact with male animals by catching the geldings with an *uurga* or collecting firewood with the oxen. Just as the herders align with quite specific gender roles, herd animals do too within the social structure of the herd. Milking is exclusively the task of women because women are thought to be more deft with their hands and attuned to the cows, whereas the men catch the horses and handle the oxen because these animals are big and powerful and require someone with strength to restrain them. If there is a shortage of labour, women (and to a lesser extent men) do engage in most tasks. The only tasks women do not engage in is in catching and breaking in the horses, or in slaughtering an animal (see Chapter 6 for horse-breaking and Chapter 10 for slaughtering). During the Soviet era, Choijo exclusively herded horses and he successfully managed a large herd. I asked Choijo if he thought herding horses was easy work and he replied "yes, because I had already learnt horse herding [from my father] and horse herding is a task for men." He added, "When I herded a lot of horses, I didn't drink a lot. I would come home, Naraa made the dinner, [and] then next morning I would go to look after the horses again. That was my work."

I only witnessed work roles becoming problematic once the men became older. Choijo, for instance, no longer had the same strength as his sons for handling horses and oxen, which meant that he had become more limited in the herding tasks that he was capable of doing, whereas Naraa was still able to continue to milk and carry out household tasks. This meant that for much of the summer, Choijo would socialise with neighbouring men of his age, and they would remain drunk on fermented mare's milk throughout much of the milking season. Choijo's main role had become the minding of the younger grandchildren, or giving occasional advice on herding activities. Older men are respected as the heads of the family encampment, but effectively they have the least work to participate in.

Gender segregation extends to the tools used for specific tasks. Tools for men are generally related to riding and traction-related activities, whereas women's tools are related to milking or cooking activities. As Bayarsaikhan (2006: 185) proudly states, "a beautiful 'uurga' is one of the adornments of a Mongolian man." The lasso-pole (*uurga*), which is only used for catching horses, is used exclusively by men, whereas the wooden ladle (*tsatsal*) used for sprinkling milk to the four winds each morning is used only by women. The woman of the family is usually the first to rise. She lights the fire and makes the milk tea that will be drunk throughout the day and then sprinkles a portion of this tea to the winds. The association of women with milk is clear in the lyrics of a herding song that was sung at a wedding I attended:

> I am a milkmaid's son (*Saalgin eejiin khuu bilee bi*),
> I was born into my Mother's milky breast (*Saalgin negerin n töroi bilee*).

Another herding song, which is sung on most social occasions, is "A Herder's Fate" (*Malchyn Zaya*), which opens with lyrics of a woman pouring milk with her arm extended to aerate the milk as it heats on the stove to make salty milk tea. The morning sunlight extends into the *ger* through the calotte (*toono*) in the ceiling of the *ger*, casting light through the milk as it is poured:

> Pouring the milk with the rays of the sun (*Togootoi cuugee nartai deverkhed*),
> While aerating the milk, it splashes on the ceiling (*Toonoo myalaan satrakh örlöö*).

Moving to Different Pastures

Once new grass shoots have sprouted up through the ground from the first rains, changing the colour of the valley to green, and by the time the lambs and kids have begun grazing on these new shoots, the herding families in the valley begin the move to summer pastures. Herders do not change locations for their own personal comfort or to make their own existence easier. The effort required to move the entire household and the herd animals is considerable, being one to two full days' work. It is largely on the basis of the herds' well-being that they migrate to new pastures.[2]

The same four seasonal pastures may be used year after year, particularly the winter camp. A winter encampment is the most established of the seasonal encampments, with timber pens, shelters, corrals, and storage huts. A summer encampment consists of semi-permanent pens and corrals (see a description of summer and winter encampments in Chapter 3). Naraa informed me that they had resided in the same summer location for the past thirty years but had moved the winter site a couple of years previously when the mountainside nearby had caught on fire, leaving meager winter pastures for the herd animals. The spring and autumn camps are more temporary sites that are occupied for only a month or so, and the herding family builds interim pens and corrals for the herd animals.

The herding families I encountered in the Khangai Mountains generally undertake four nomadic migrations in one year to coincide with the four seasons, with separate spring, summer, autumn, and winter camps. There are a fewer number of migrations here than there are in many other parts of Mongolia because the Khangai has particularly lush pastures and reliable access to water sources. Bold (2001) explains the migrations between the winter–spring site (*ötög-buus*) and the summer–autumn pasturage (*bilcheer-us*) as being the longest and the most important migrations (see Figure 8.2). Depending upon weather conditions and the state of the pasture, herders may move to an *otor* camp, which is a temporary site where the herd animals

[2] See Moving Encampments: Part 2. Moving with the Herd Animals, available at www.cambridge.org/9781107000902.

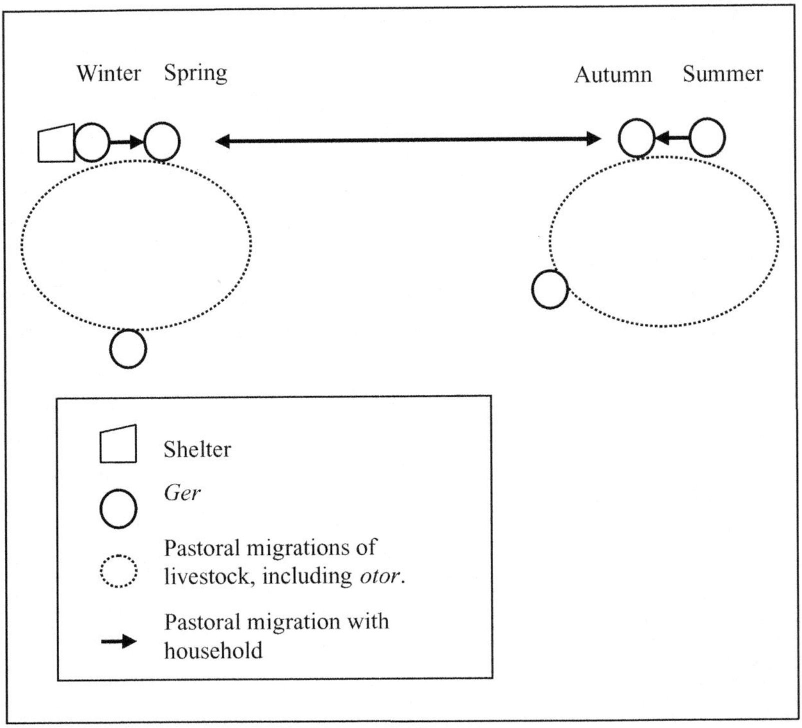

Figure 8.2. Migratory movements between seasonal pastures and location of encampments. (Source: Adapted from a diagram in Bold 2001.)

graze on pasture that is separate from regular seasonal pastures. For the purposes of *otor*, or shorter periods like the spring or autumn sites, herding families live within a smaller *ger* and store some of their possessions in a winter storage hut. One herding family may move on their own with just one herd, usually the sheep and goats.

Herding families choose where their *ger* will be situated on the basis of the best position in relation to the needs of the herd animals.[3] In Dogsomjav's encampment, Dogsomjav remains at the winter site, tending to the cows and calves, while Ochero's family move with the sheep and goats and their newborn young to a temporary spring encampment situated approximately one kilometre from the winter shelters. Ochero's family moves the sheep and goat herd to prevent the lambs and kids from becoming too hot with the warmer

[3] There are other factors, such as those mentioned by Erdenebataar (1996): type of livestock; climatic and geo-ecological characteristics; condition of pasture; availability of transport, and water or salt sources; however, these factors are principally with the movement of the animals in mind.

spring weather conditions. The separation of herds is a common strategy for herding families, used to divide the tasks between families within the same encampment.

Herders within each broad river valley in the Khangai Mountains coordinate their nomadic movements with one another as they move up or down the valley. The family decides upon an auspicious date to move that is based on the Mongolian calendar. All families then move within a few days of one another. When I asked why all the families move at the same time, a herder gave me a practical response: If only one herding family remains in one portion of the valley, the wolves in that area focus all their attention on the herd animals of that one remaining family. Before a seasonal move, the herds have changed their grazing routines and have started searching for new pastures.

On the day of the move, the adult herd animals seem to know they are migrating and respond accordingly. The adult herd animals know the routine of the move because they have migrated up and down the valley between the main summer and winter pastures in a cyclical routine. During the migration, the animals divide off into different groups of their own accord: the stallion with the mares and foals; the geldings as another group; the cattle, including the cows with their calves; and the sheep and goats with their lambs and kids.

In the text that follows, I describe the migration of the herd animals from the spring pasture to the summer one, which is approximately twenty-five to thirty kilometres up a broad river valley (for detailed definitions of the different kinds of migratory pastoralism practiced by the herders of Inner Asia, see Khazanov 1994). Dogsomjav and I herded the animals, while the rest of the extended family and their belongings were taken by a friend's truck[4] and on ox carts (movement by ox cart will be described later in this chapter). At the start of the move, the herd animals begin at a fast pace in their anticipation of new pastures. The horses are not overtly instructed or guided by the herders along the way but trot on ahead of their own accord. They are considerably faster than the other herd animals, so they are free to stop and graze along the way.

As can be seen from the following field excerpt, Dogsomjav clearly thought the bull ("*Khurnee*") was anticipating the arrival at the summer site, which meant he caused problems for her along the way. I also assisted a neighbouring herder with herding his animals to the summer site and asked

[4] The cost of petrol for moving by truck is high. Dogsomjav was intending to exchange two sheep for the use of the truck, but in this instance I assisted the family by paying for the truck with money (the time and effort invested in raising a sheep outweighed the cost of the petrol in cash, but a live sheep cannot be divided into portions).

him whether he thought the animals remembered the summer pastures; he confirmed that he thought they do.

> I was amazed that the horses did not need any guidance whatsoever and independently made their way up the valley to the summer pastures. Dogsomjav concentrated on guiding the female cattle with young that had less strength and endurance, while I concentrated on herding the sheep and goats with their young. The calves had the hardest time coping with the long journey and a few lagged behind with their attentive mothers.[5] Initially the bull, Khurnee, made herding the cattle difficult because he kept turning back towards the spring encampment, chasing the cows and lying down on the ground, obstructing progress. When the bull rubbed his head and neck in the dirt and commenced bellowing, Dogsomjav said the bull was preparing himself for battle with the neighbouring bull, who would soon be in close proximity at the summer site.... One woman brought food and tea for us as we passed by her encampment. It is a herding tradition to help any passersby by offering food and tea and holding the horses for the herder while they take a break. Two encampments came to assist with the herding for short distances, which was greatly appreciated, as it became tiring constantly calling, whistling, and signalling to direct the animals. Women and young children came out as we passed their encampments to keep their animals out of the way, especially the young calves. When the young animals would lie down, Dogsomjav would stop for a break. Luckily we had cloud cover in the afternoon, which meant it was not too hot for the animals. When we arrived, the rest of the family helped to drive the calves into their newly erected pen, while the lambs and kids were left to graze until nightfall. (Field note excerpt: Moving to Dogsomjav's summer encampment, May 21, 2005)

SUMMER ACTIVITY

For most of the year in Mongolia the vegetation stands as dried stalks with no new growth. Once grazed, the grass does not grow back and the herd requires new pasture. It is only during the short summer season that the landscape thrives with a spurt of new growth; this is therefore called the "resurrection period" (Tomikawa 2006). The ground becomes muddy underfoot as the upper layers of soil thaw; instead of snow, it is warm enough for rain to fall. With the rain and the heat of summer come the insects that hatch out of nearby rivers and grassland. The flies are numerous and persistent, attracted

[5] The morning after we arrived at the summer encampment, a calf was born prematurely. Lhagva informed me that the cow may have given birth as a result of the stress of the journey the previous day.

to people and herd animals alike. Horses repetitively nod their heads in unison as they walk and swish their tails to deflect the flies that gather in the hundreds around any one individual. The herd animals have lost their shaggy winter coats, and instead their coats develop a healthy shine (see the insert of summer photographs, shown in Plates 9–16).

Summer is the season when herders in the Khangai Mountains scythe hay to provide supplementary feed for the animals during the cold of winter. Herders use the long grass from their reserve pastures near their winter site to cut hay. This means that all the men from the encampment ride by horseback on excursions from the summer site to the winter pastures to scythe the hay. Young herders, like Saikhanaa, or older herders, like Choijo, aerate the hay with pitchforks, and then gather the rows of scythed hay into neat, domed haystacks for transportation by ox cart to the winter site. The hay is stored on the roofs of the winter shelters so that there is some feed for weak or sick animals during the winter and spring.

For herders, a summer's day is long and there is plenty of work but summer is also a social time, with plenty of food. The central fire from the *ger* is moved outside during summer, as it becomes unbearably hot from the heat of the fire inside a *ger* when the outside temperature may be 30° C. Herders visit neighbouring families to partake in meals outdoors, or to drink the abundant fermented mare's milk (*airag*). The whole structure of the extended family encampment changes during the summer, as older children come home from living with relatives, or from boarding in the nearest town, where for nine months of the year they attend school. Town-dwelling relatives may send their children to stay in the countryside and they are expected to assist with work activities in exchange for sleeping and eating within the encampment.

Women make meals between milking the animals and making dairy products. In summer, the family may eat milk tea and fried *buuzig* for breakfast; dairy products as snacks throughout the day; and one large, cooked meal of meat with flour noodles or rice in the evening after most of the work with the animals has been completed. Herders told me repeatedly that the tradition is to eat fatty meat over the colder months but then it is necessary to cleanse the system by eating mainly dairy products during the summer months. According to the Nestorian monk, William of Rubrick, in his thirteenth-century account of Mongolians, they drank only *airag* and ate dairy products in the summer and meat was only eaten in exceptional circumstances (Bold 2001). At the beginning of the twentieth century the caloric intake of milk products accounted for 55 percent, vegetable foods 24 percent, and meat for only 20 percent (Khazanov 1994). Now the proportion of meat products consumed would easily be the highest proportion.

A large proportion of the women's work involves making dairy products from the milk that is obtained each day. Women, in particular, are bound to the encampment to milk the animals, as they have to milk according to a daily routine; otherwise, the herd animals do not produce a consistent amount of milk, which is required to feed the family throughout the rest of the year. If any one woman is absent from the milking, another woman from the encampment needs to replace her to milk the cows she usually milks.

The herding women are required to process any milk the herd animals produce almost immediately because there is a lack of refrigeration and to ensure no milk is wasted. Milk from the cows makes up the greatest proportion of the various kinds of dairy product. These dairy products are made by boiling; adding bacteria from yoghurt; adding sugar, or flour; and by drying curd as preservation techniques.[6] Three different dairy products can be attained from different proportions added to yoghurt, or milk: *byaslag*, which is a cheese made from warm milk with a small portion of yoghurt; *aarts*, which is curd made from boiled yoghurt with a portion of milk; and *aaruul*, which is made from *aarts* after the liquid has been removed and the curd has been pressed overnight. It is then sliced into thin slivers and dried (either around the ceiling from the heat of the fire, or on the roof of the *ger* by the sun). Cream (*öröm*) is made from boiling milk and skimming the upper layer off after the milk has been left overnight. *Sarlag* milk is thicker than Mongolian-cow milk and yellow in colour after it is left to congeal overnight. The cream from the *sarlag* is over an inch thick, or in Mongolian measurement terms it is two-finger widths in thickness. Any cream that is not eaten immediately is stored in sheep stomach to preserve it until winter and then make it into *shar tos* (yellow butter).

Mare's milk is not boiled but fermented in a calf-hide container (*khukhuur*) by churning the milk between 800 and 1,000 times. The milk is then left to ferment overnight to make *airag*. All ages assist to churn *airag* and all ages drink this alcoholic beverage, including babies. A small portion of the *airag* culture remains in the family's *khukhuur* and then more mare's milk is added after each milking bout throughout the day. Mare's milk contains valuable nutrients that are otherwise not present in the herding diet and is considered to have antibiotic properties that protect against bacterial infections. Mare's milk, unlike the milk from cows, sheep, or goats, does not require boiling, as Mongolian horses do not carry diseases such as tuberculosis, brucellosis, and inflammation of the udder (Indra 2003). Mongolians view both horse meat and milk as "clean," in comparison with the meat or

[6] See the Glossary of Mongolian Dairy Products for the different types of dairy products made by herding women and the basic techniques for making them.

milk of the other kinds of herd animal. When I first drank *airag*, the herding family warned me that it would cause me to have diarrhoea at first but even this is thought to be beneficial as a means of "flushing out the system" (see Chapter 7 for the social importance of *airag* in Mongolian society). Milk from mares, cows, nannies, and ewes may be distilled further to make *nermel*, a clear alcohol that is only offered to visitors or drunk on special occasions.

Mongolian herders particularly praise hard work. Naraa emphasised how hard she used to work when she was younger by the adeptness of her milking skills.

> When I was young, I milked sixty or seventy mares in one hour alone. When I milked, three men would handle the foals for me. At that time mares had a lot of milk...when I milked three mares the bucket would be full. At that time mare's udders were easier to milk, like the Dun mare's [udders]. I also milked fifteen cows every morning and evening, because then we were very busy, but now we are more relaxed with less work. (Naraa, summer encampment, August 17, 2005)

A common insult amongst herders is to call people or individual herd animals "lazy" (*zalkhuu*). Herders assumed that people living in cities lead a lazy, idle existence. I commented to Lhagva that I thought Mongolian herding women worked hard; she agreed, saying "mostly countryside women are very hard-working because we have to look after our babies, the animals, make food and keep the home clean." She went on to say that "men are lazier than women. Mongolian women do almost everything. Women are very hard working but men are just here for socialising." According to Bold (2001: 64), "expenditure of labour differs depending on the season and the separate types of activity. The image of a shepherd sitting near his grazing sheep, chatting the day away or snoozing, does in fact represent nomadic livestock keeping work, and such images are characteristic." This stereotype of the herdsman may be applied to the work carried out by men, but the image of laziness could less readily be applied to the work of women.

In the text that follows, I depict the daily activities with the herd animals in summer with the herder, Saikhanaa, as an example. In 2005, Saikhanaa was fifteen (fourteen in Western terms) and a particularly knowledgeable, active, hard-working herder for his age. He definitely would not be described by others as lazy. I single out Saikhanaa, because he participates both in milking activities (because he is still young – he will cease to do so as an adult male) and also in male-oriented activities, such as assisting to catch the geldings for the day's herding, herding the sheep and goats, or catching the foals for the tethering lines (*zel*). Unlike the rest of the children and teenagers from Choijo's encampment, Saikhanaa did not attend school during the year, as he was expected to assist his mother and work for Choijo and Naraa (see The Prologue).

A Summer's Day in the Life of Saikhanaa

In June, dawn comes as early as five in the morning in Mongolia. Saikhanaa and his mother, Khorlo, were the first to rise in the encampment. They released the sheep from the pen so that the herd could begin to graze, and they tied up the dogs that had been left to roam during the night, guarding the encampment from wolves. By June, the lambs and kids are no longer segregated from the adults because they can feed on grass for themselves and are capable of following the rest of the herd out to pasture each day. One gelding would be hobbled nearby during the night, so that Saikhanaa could vault onto him bareback and ride off to find the cattle herd. The herd was harder to find on some days than others. By late summer the cattle were moving further from the encampment towards the autumn pastures, so Saikhanaa would spend up to an hour retrieving them. The cattle herd arrived at the milking area, already feeling the heat with their thick coats. The *sarlag* would be panting like dogs, their long purple tongues hanging from their mouths.

Once the women arrived with their wooden pails to milk their cows, Saikhanaa released the calves one by one to feed briefly from their mothers. As each cow was needed for milking, Saikhanaa would fetch the cow and drive her near the calf pen, and then signal for her to stop so that he could hobble her front legs. He would call the name of the calf at the gate of the pen and release the calf to run and suckle from its mother. While that calf was suckling he would hobble another cow, or release another calf. Once the calf had ample opportunity to suckle on all four teats, Saikhanaa removed the calf so that one of the women could begin milking her cow immediately and the calf would be directed back into the pen.[7]

Cows and calves are recognised as belonging to a particular family within the encampment, and it is those family members who milk their own cows. If the "owner" (usually an extended family member) is absent, then one of the women from the encampment will milk the cow for her. Children or visitors (including me) would assist with the job of "hobbling" but Saikhanaa tended to prepare Naraa's cows in particular because she is the matriarch of the family and has the most cows. When the cow had been milked, Saikhanaa would release the calf from the pen to suckle from its mother for a second time. The

[7] Each encampment has slightly different techniques for milking cows, and this is dependent upon whether the cattle consist of *sarlag* or Mongolian cows. In Dogsomjav's encampment, the calves were tied to the outside of the pen while the cow was being milked, rather than returning them back inside the pen. Other families with fewer cows to milk did not bother to hobble the cows at all and would let the cow lick her calf while being milked.

calf is left beside its mother until it ceases nursing and is then returned to the pen. The following is a field note extract of the roles between the person doing the hobbling (*unee evelgegch*) and the women who milk (*saal'chin*)[8]:

> The social hierarchy during milking is somewhat different from the hierarchy within the family *ger*. The women who milk the cows are clearly dominant over anyone with the job of hobbling, as the women instruct when the calf should stop sucking and therefore when the cow's teats are ready for milking. The women also instruct the hobbler about which cows to hobble for them to milk (perhaps this is why I never saw men hobbling and handling the calves, only teenage boys). The hobbler is expected to wait while the calf sucks, until the milker tells him whether he can take the calf away from its mother's teats or not. Milking is one of the most important activities in herding life and is taken seriously. The atmosphere is quite different while working during milking compared to the more relaxed, sociable atmosphere inside the family *ger*. Younger children are not expected to participate in the cow milking activities until they are big enough to handle the strength of the calves, even though they already have a good knowledge of the individual cows and calves and of the milking process. It is the hobbler's job to keep the milking running smoothly, so that the cows are ready to be milked as soon as the milker finishes the previous cow, so that she can move quickly from one cow to the next. (Field note excerpt: Choijo's summer encampment, August 2, 2005)

Choijo's encampment had a large number of cows to milk, with twenty-eight calves in one pen and seventeen *byaruu* in another. This would mean it would take up to two hours to finish the milking process in the morning and the whole process would be repeated again in the evening for one and a half hours (it took less time in the evening, as the cows produce less milk during the second milking of the day). Saikhanaa knew each one of the calves and *byaruu* and to which mother they belonged.[9] After all of the morning's milking had been completed, the calves were briefly returned to the pens and the cows released from their hobbles. Saikhanaa would then drive the herd out to pasture. The calves were only released from their pen once the rest of the herd was out of sight. I asked Saikhanaa, "what was the most difficult task for you to learn?" He replied, "hobbling the cows and being able to recognise

[8] The term "milkmaid" in English implies a servant, or a woman of low status, whereas milking in Mongolia does not have the same connotations. I refer to the person who milks the cows as a "milker," instead of the old-fashioned, gender-biased term of "milkmaid." The same premise is used when I refer to a person herding as a "herder," rather than as a "herdsman," because many of the herders are women and children.

[9] See Saikhanaa and the Calves, available at www.cambridge.org/9781107000902.

which cow belonged to each calf…that was difficult." (Saikhanaa, summer encampment, August 18, 2005)

As the women finished the morning's cow milking, the men retrieved the entire horse herd from where they had been grazing and drove them into a large corral to catch geldings for the day's herding activities. Any horses that were showing signs of illness were caught for treatment, and the men selected geldings for riding during the next day or two. It was Saikhanaa's job to retrieve the correct halters and bridles from the *ger* and hand them to the men while they caught the horses. Saikhanaa would not usually catch the horses with the *uurga* because, on foot, it requires men who are stronger and can apply all their weight to the end of the pole. If a horse has never been caught (*khangal*), four or even five men will hold onto the *uurga* to restrain the horse. The herder pulls "at the moment the forefeet of the caught horse are at [sic] air. Horsemen say that when a horse is caught in such a way, the strength of the horse is as light as a foal's" (Bayarsaikhan 2006: 187). Usually, however, the horse has been caught enough to stop as soon as the loop of the *uurga* is around its neck. The horse calmly stands while the herder approaches with a bridle in his raised hand, and then places the bit between the horse's teeth. The horse is then led away to be saddled up and tied to the hitching post to stand throughout the day until required for herding or other tasks.

The next task of the day for the men and boys of the encampment is to catch the foals and two-year-olds by hand. Mares that have not given birth that summer may continue to feed their young from the previous year. These two-year-olds are called *khusran daaga*. Each *khusran* is caught and haltered and then placed in an allocated position along a tethering line. In Choijo's encampment the tethering lines were set out along five lines, each categorised according to a stallion and the respective mares and foals that were part of the stallion's band. The foals are easier to capture than the two-year-olds because they remain close to their mothers. After a month or so the foals are so used to being captured that they will remain sitting on the ground when the herder approaches, or just stand motionless while the herder slips one arm over the foal's neck and with the other places a halter over the foal's head. Once all young horses are caught, the herd is left for two to three hours to graze near the tethering lines until the mares have produced enough milk for extracting a small amount.

It is Saikhanaa's job, and that of other males, to handle the foals or *daaga* during milking times. Mares are milked six to eight times a day because, unlike cows, they only produce a small amount of milk before the udder is full. The mare usually stands near her foal to allow it to feed, but with a herd of over 100 horses, Saikhanaa requires a sound knowledge of each individual in order to match the foals with their mothers. The foals are prevented from feeding from their mothers between milking times by being tethered. If a mare still tends to approach her foal to allow it to nurse, one of the herding women sews a cloth flap onto the halter, so that the foal's mouth is covered by the cloth when it attempts to suckle. Each time a mare is to be milked, the foal is guided to its mother and allowed to nurse for a few seconds until one of the women arrives to milk the mare. While the mare is milked, the foal is held parallel to its mother and then allowed to feed briefly for a few seconds after milking. Both Saikhanaa and the woman milking make encouraging noises to the mare, mimicking the sucking noise of the foal as it nurses, or yell "*guurii*," the command to release milk. It is during this process that the herders begin to know the horses on a one-to-one, individual basis, the female herders through milking and the male herders through handling the young horses.

There is less activity in the middle of the day, while the herd animals rest from grazing. The sheep and goat herd is driven down to the river to drink, while the horses and cattle make their own way down to the river in long processions. Herders are very aware of the amount of water the herd animals require, especially in summer when they require more water. Horses and oxen that are being worked become tired and fatigued if they are thirsty. When the animals rest, so too do the herders, lying down on their beds in the *ger* to escape from the heat. The canvas sides of the *ger* are tucked up to encourage a breeze to flow through the hot tent.

Throughout the day, Saikhanaa assists with the mare milking, in between watching the sheep and goat herd.[10] By summer, the sheep and goats require less attention because the pasture is abundant so they do not stray as far from the encampment. The lambs and kids have also grown large enough to follow the herd, are capable of feeding themselves by grazing, and are less vulnerable to wolf predation. Saikhanaa would spend most of his time out herding during the winter and spring, but by summer one of the men would observe the herd through binoculars and directed Saikhanaa to retrieve them only if they were straying too far across the mountainside. If they entered the forest, where the risk of attack by wolves is more likely, Saikhanaa would stop the task at hand, vault onto the gelding standing at the hitching post, and redirect the herd by rounding up the stragglers and directing lead animals through vocalisations. Sometimes the final bout of mare milking would not

[10] Here I have left out Saikhanaa's involvement in training racehorses for the *Naadam* season and his participation in many *Naadam* events throughout the summer (see Chapter 7). Sheep and goats are also milked for approximately one month in the middle of the day, as they are herded back to the river valley to drink water and to rest from grazing in the heat of the midday sun.

be until ten o'clock at night. After each mare was milked, her foal would be released from its halter, allowing the entire herd to move down to the river to drink. At dusk, Saikhanaa's final task for the day would be to ride to the sheep and goat herd, perhaps one or two kilometres away on the nearby mountainside, and direct them back down to the encampment and inside the pen for the night. It is evident from these descriptions that a herder's day, particularly during the summer, is entirely structured around the routine of the herd animals.

Moving to a New Encampment

By the end of summer the leaves have begun to change colour, the first frosts begin, and the summer pastures have been grazed to stubble. For the move from the spring to the summer encampment just detailed, I provided a description of the movement of the herd animals to the summer pastures. For the move from summer to autumn pastures, I will illustrate how possessions are moved by ox carts to the next site.[11] Many herders use trucks to move longer distances but for shorter moves to interim camps, such as the spring and autumn encampments, herders use handmade, wooden ox carts. According to herders in Bulgan Aimag, oxen are still crucial in areas where the terrain is too steep for motorised transport. The structure of these carts is very similar to the structure of carts depicted in Bronze Age rock art found in Mongolia (see figures in Sorokon 1990). In early September, Choijo's encampment and all the surrounding encampments prepare to leave for their autumn site. Movement to the autumn site is stimulated by a sharp drop in temperature and by the fact that the herds begin moving further and further afield to find pasture that has not already been grazed. The move from the summer to the autumn pastures is usually much shorter than the previous move from the spring to the summer encampment (see Figure 8.2). On the day that the household possessions were to be moved, Choijo's sons and neighbouring herders worked together to herd all the oxen into a large corral and catch them with a cattle-hide lasso (*buguil*).[12] A herder threads a rope through the nostrils of each ox, so that it is easier to control while pulling the ox cart.

Movement to a seasonal encampment is a communal activity and many of the neighbouring herders, particularly the men, ride over to assist the family that is moving. This is then reciprocated within the next few days when it

[11] See Moving Encampments: Part 1. Moving Encampments by Ox, available at www.cambridge.org/9781107000902.

[12] Choijo's encampment had fourteen oxen in 2005. The oxen and accompanying ox carts were borrowed by neighbouring herders to move encampments. It took nine oxen to move two families and the contents of their respective *ger*.

is the neighbouring encampment's turn to move. Each *ger* is quickly dismantled by the men, and then canvas, poles, and felt are loaded onto an ox cart. Meanwhile, the women pack up the contents of the *ger*. Saikhanaa would assist the men packing up, but it was his job to lay out the harness for the ox cart and hitch the oxen to the carts once they were loaded with household possessions. A calm lead ox leads the ox train, followed by the rest of the oxen with their loaded ox carts. As the first ox departs, the next ox feels the tug on the rope through his nostrils, and so moves off. Sometimes herders ride beside the oxen on horseback, so that they have the mobility to spur on any individual ox that is lagging behind, or to retrieve any of the possessions that may fall from one of the carts. Saikhanaa rode or walked beside the lead ox, spurring it on with a stick and vocalising to the rest of the oxen to "move off" (*khudg*), or "halt" (*drrr*). The following is a field note extract detailing the move from Choijo's summer to the autumn encampment.

> Saikhanaa led the string of nine oxen with their ox carts as they slowly left to make their way to the autumn location with all the household possessions on the ox carts. This move was a relatively short distance of approximately ten kilometres. The cattle have been anticipating the move to the autumn pasture, as the herd kept grazing near the autumn pastures for a week before the move. As soon as we arrived at the autumn site everyone began assisting with setting up each *ger*. By nightfall the women had arranged the inside of the *ger* in exactly the same layout as it had been in the previous summer camp. In the evening the oxen were all taken down to the river to drink by riding one and leading two on either side. The oxen that were used to move Choijo's encampment's possessions were then used to move Sukhee's encampment [Naraa's brother's extended family]. (Field note excerpt: The move to Choijo's autumn encampment, September 6, 2005)

When moving to a seasonal encampment, the herd animals are tended to before the needs of the herding family. The family then gathers to recognise the site as "home." The woman of the household sprinkles milk tea to "the four winds," and then cups of salty milk tea are drunk, because tea is always served first before any other food or beverage. The *ger* and any pens the animals may require are erected by using the same process with which they are dismantled. It takes the next few days for the herd animals to get used to a new routine, with different locations for the calf pens and tethering lines for the foals, but this also means that herders can take advantage of the disruption of routine to change milking times to align with dawn and dusk.

At the autumn site, members of Choijo's extended family were situated near other relatives' encampments. The pasture was long and lush, so the herds had plenty of pasture to graze upon. Amaraa's parents had moved a week previously (on August 28, 2005) and had their autumn encampment 100 to 200 metres from Choijo's encampment. Two other encampments, including Naraa's brother's extended family, moved the following day and set up their *ger* adjacent to Choijo's encampment. This meant that there were twelve *ger* all in close proximity to one another and they used the same autumn pastures for their herds. This was feasible because they were in that location for only a month or so, before moving on to their respective winter sites.

AUTUMN ACTIVITY

Many Mongolians said to me that autumn is the best season because the days are still warm, there are fewer flies and biting insects than in summer, the weather is quite constant without the high winds and snowstorms of spring, and herders are not as busy because the young animals are strong enough to feed for themselves and there is less milking to be done. Nights become colder again with frost on the ground during the morning cow milking. The women huddle against the cows' bodies for warmth as they milk. Amaraa pointed out to me, during mare milking, that when the mares start to "get angry" (kicking and biting) at their two-year-old offspring it is a sign that autumn is coming, as this is when the two-year-olds are weaned by the mares; this comment was soon affirmed by a sudden drop in temperature. The leaves on the larch trees (*Larix siberica*), followed by those on the black pines (*Pinus sylvestris*), begin to change colour. Once autumn has set in, the valley no longer displays the vibrant greens of summer but changes to the muted hues of autumn. Autumn is essentially a time to prepare for the long, extremely cold winter. The herd animals begin to grow their thick winter coats and herders don their lined autumn coats (or *deel*; see the insert of autumn photographs, shown in Plates 17–24).

In autumn, Ochero moves to the same location as the spring site to milk mares from mid-October to mid-November. Unlike members of Choijo's encampment, where they milk mares from June until October each year, members of Dogsomjav's encampment only milk for one month, just long enough to collect the milk needed to make the distilled Mongolian vodka (*nermel*) for consumption during the *Tsagaan Sar* celebrations in winter (see Figure 8.1). Dogsomjav told me that they did not milk throughout the summer in 2005 because they did not have many women to milk the mares and there were not many foals born, so there were few mares to milk. It was for these same reasons that Dogsomjav's encampment combined their mare milking with that of the neighbouring encampment. One of the neighbouring herders, Nasaa, would ride over every two hours to assist Ochero with handling the foals, while Boloro milked each mare.

Whenever Dogsomjav herds the sheep and goats in autumn, she searches for medicinal plants and collects them for use throughout the rest of the year (refer to Chapter 9). In preparation for winter, the herding families repair the shelters for the herd animals; they ensure that any gaps in the birch pole walls are sealed with animal dung, and they fell fresh birch poles to repair pens or larger corrals. As an example of preparing the herd animals for the long winter ahead, Lhagva told me that they give the animals handfuls of salt in autumn "so that the animals stay warm and strong. If people give salt to the animals, then they are not as likely to become cold in the wintertime" (Lhagva, autumn encampment, October 21, 2005).

Herders condition the herd animals to survive the frigid temperatures of winter (Tumerjav 2003). Herders direct the sheep and goats to graze at cooler pasturing locations. October is the time for hardening the geldings (*mor' soijakh*) so that they can be ridden periodically throughout the winter. For example, in Dogsomjav's encampment, the geldings were round-bellied animals in good condition from grazing on abundant pasture all summer. On October 10, Ochero tied eight geldings at a time to the hitching post and left them there for two nights. He would only allow them to drink once during the two days, leading them down to the river in the evening.

Herders cull their herds at the end of each autumn. By this time of year the temperatures have dropped to below freezing. Each family chooses animals for slaughter that are less likely to survive through the winter because they have some form of weakness. After an animal is slaughtered, everyone within the encampment partakes in the bowl of innards, which require immediate consumption. The rest of the meat from the animal is hung inside a storage hut to dry and to be consumed over the long winter. Any extra innards are distributed amongst neighbouring herding families. Herders also slaughter and process animals to provide meat for extended family members who live away from the encampment during the winter. Khazanov (1994) refers to this distribution of food amongst other members of the community as "balanced reciprocity," which is an important principle of Mongolian exchange. Each herding family generally eats the meat from their own animals. This extends to "absentee owners," who will be given the meat from any of their own animals that are culled.[13]

[13] Fernández-Giménez (1999) uses the term "absentee herders" to refer to herders that own livestock but are absent from the herding encampment. The majority of the absentee owners are kin or acquaintances of the herders.

Dogsomjav had an old cow, Oodon, who originated from "the company" or collective (*negdel*). Oodon gave birth to the calf Menget in the spring and had produced six calves since the family had acquired her. During the move from the summer to autumn pasture I noticed that she had become slower and was obviously more fatigued than the other cows. The family fed her some supplementary feed but she had lost many of her teeth. On an auspicious day, November 10, 2005, Oodon was part of the autumn cull. If an animal is young, healthy, and producing offspring, then the animal will not be selected for slaughter. It is only those animals that are older, not producing young, or have some ailment that are culled for food.

If the herd is not healthy or productive, then the herders cannot survive their herding existence, as meat and dairy products are their primary food source. The extreme conditions of winter in Mongolia mean that there is strong selection pressure from environmental factors. If an animal is unhealthy or not as fit as others, then the individual is unlikely to survive the winter.

> These life conditions mean that they need to co-work in order to survive this lifestyle. That's why this evolutionary process progresses relatively smoothly because the harsh conditions help them both to understand each other as co-workers in collaboration. This [herding] life cannot exist without this co-understanding and co-working for survival; it's been very helpful for their understanding of each other. (Enkhtuvshin, Ulaanbaatar, November 2, 2005; spoken in English)

WINTER ACTIVITY

With the onset of winter, snow and ice cover the ground and the days become short. Both herders and herd animals adopt a state of semi-hibernation. Life slows down as metabolisms slow to conserve energy. Herders do not move the horses faster than a slow jog so that they retain enough energy to survive through the long winter season, in which temperatures remain at -15° C or below. Naraa told me (as the first to rise in her family *ger* in the morning during the long winter months) that she gets up at 8 A.M. and goes to bed around 8 P.M. after the animals have been put in the pens and shelters and it is dark. This is a lot more sleep than she gets during the long days of summer (see the insert of winter photographs, shown in Plates 25–32).

In winter there are few visitors and the encampment is reduced to the core herding families, with children who are not yet of school age. The herding family remains inside the warm cocoon of their *ger* during much of the day.

Each *ger* is insulated by three layers of felt (made from sheep wool) and by animal dung, which is collected and poured around the edges of the outer canvas to ensure that no draft can enter at ground level. Likewise, herders ensure that the shelters are insulated by layers of dried sheep and goat dung (*utug buuts*). Lhagva emphasised to me the importance of maintaining warm shelter for the animals. In winter "we have to look after the animals very carefully, so we need to give them hay and make sure they don't become too tired. We provide a comfortable and warm shelter ... it's very important for the animals" (Lhagva, autumn encampment, October 21, 2005). The herd animals do not tend to stray far from the encampment because the winter reserve pasture is usually in close proximity. Daily interaction is more towards the herd as a whole, not towards individuals, as there is little herder–herd animal interaction in winter without the daily milking routine, or the need to nurture young animals. The importance of warm shelters and skilled herding during the winter was emphasised by Choijo when he said "look at my horses; it looks like we have more than others [in the valley]. This is because in the winter the animals survive, so the horses are my reward [for good work]" (Choijo, summer encampment, August 17, 2005).

CONCLUSION

Life in Mongolia occurs in cyclical patterns, in which events are repeated from one year to the next, inexorably moving from one season to the next (see the seasonal calendar shown in Figure 8.1). Every herder knows that spring is the time for haircutting and combing, assisting animals in birth, and nurturing young; summer is the time for milking, making hay, socialising, and attending *Naadam* festivals; autumn is the time for castrating horses, gathering medicinal herbs, and culling the herds for the winter ahead; winter is the time of rest and retiring into the warmth, preserving energy, and celebrating the coming year with the White Month (*Tsagaan Sar*) celebrations. In Mongolia, the four seasons are pronounced. With each season the family moves to different pastures for the benefit of the herd animals. The herd animals' routine changes after each move, so therefore the work the herders engage in also changes. Herders rise at dawn for the first milking tasks of the day and go to bed soon after the sheep and goats are herded in to the pen just after dusk. Herders sleep when the herd animals sleep and are active when the herd animals are active. Work roles vary depending upon the gender and age of the herder, and this coincides with the gender and age groups of the herd animals themselves. I would be unable to summarise the parallel lives of herder and herd animal more succinctly than Tim Ingold has done: "there is thus a sense in which people and their domestic animals grow older together, and in which their respective life-histories are intertwined as mutually constitutive strands of a single process" (Ingold 2000: 86). As my Mongolian friend Tuvshin put it, herders and herd animals "co-work together to survive."

9

THE DOMESTIC AND THE WILD

Biology is relentlessly historical, all the way down. There is no border where evolution ends and history begins, where genes stop and environment takes up, where culture and nature submits, or vice versa. Instead, there are turtles upon turtles of naturecultures all the way down. (Haraway 2004: 2)

INTRODUCTION

To Mongolian herders, "wild animals" are those that do not enter the domestic sphere of the encampment. Herd animals are seen as part of the extended family. According to Naraa, a wild animal is afraid of humans and hides on the forested slopes, or high in the mountains where few herders would venture on horseback. A co-domestic animal is "fed by people" and nurtured as part of the co-domestic sphere (*geriin tejeever*). When I asked Naraa what she thought "to domesticate'" meant, she answered with a succinct, practical reply: "we feed them and they feed us." In fact, Mongolian herders provide only minimal supplementary feed to the herds; however, I interpret her answer to mean more broadly that a herder's role is to nurture, and that then the animals will in turn provide for the herder.

In the previous chapters I focused on herd animals and their closely intertwined lives with herders within the co-domestic sphere of the encampment. This chapter goes beyond this co-domestic realm to encounter other animals and plants that live outside this sphere, inhabiting the forest and mountain slopes and those beings that do not have regular contact with humans. Mongolian herders view wild animals and plants as harbouring great power, with a strong "spirit." Many plants are believed to be powerful as they contain healing properties used in the treatment of both human and ungulate illnesses. The wolf holds a slightly different status to other "wild" animals. Wolves are believed to be afraid of humans and mainly remain within the forest to hunt, but they occasionally transgress by venturing down into the co-domestic sphere of the encampment to prey on young livestock at night (see Figure 9.1).

Categorisation of the Wild and the Domestic

Mongolians speak of the "wild" as a separate category from the "domestic," but these concepts are not viewed as dualistic opposites; nor is it believed that wild and domestic do not engage and interact with one another. Mongolian herders have categories for such beings as humans (or people; *khün*), horses (*aduu*), dogs (*nokhoi*), and wolves (*chono*), but each exists in its own right, with a distinct society and perhaps even a "culture" of its own (although an animal's culture can be influenced by Mongolian herding culture and vice versa).

Across many human cultures, people categorise organisms at a species level, which aligns with Western scientific nomenclature traditions (Diamond 1966; Dwyer 1996; Gould, 1979). Mongolians also classify their animals on a species level: They have separate terms for the Mongolian horse (*aduu*) and Przewalski's horse (*takhi*); likewise Mongolian sheep (*khon'*) in comparison with mountain sheep (*argali*); and the Mongolian goat (*yamaa*) or the mountain goat (*yangir*). These animals are clearly recognised as distinct species and are categorised quite separately from one another but not necessarily defined according to domestic and wild. In a work that takes cues from the ideas regarding taboo of Mary Douglas and Edmund Leach, Tambiah (1969) divided the Thai Baan Phraan Muan villager's eating taboos according to domestic versus wild forest animals.[1] Mongolian herders, however, do not divide animals into those that are edible, or those that are inedible, because many are eaten for medicinal purposes, such as the dog or wolf, that are not generally consumed or thought of as food.

In Mongolia, a co-domestic animal is one that is tame through nurturing, or nourishment, thereby bringing the animal into the home (*geriin tejeever*). There is a term for wild, *zerleg*, in Mongolian (this same word is also often used to mean "crazy"). According to Simpson (1938), this term is used in essentially the same manner as it is used in English. It is my impression, however, that *zerleg* is a term used by Mongolian herders for a behavioural characteristic, rather than in terms of a separate domain that is perceived as "wildness" or "wilderness." Herders recognise that some animals can be wild (*zerleg*) and yet others can be tame (*nomkhon*) in their behaviour (see Chapter 6).

The herders I spoke with characterised wild animals as those wanting no contact with humans, or those animals that are *afraid* of humans. The old

[1] Also see Haudricourt (1969) for a division of domestic animals into dualistic opposites.

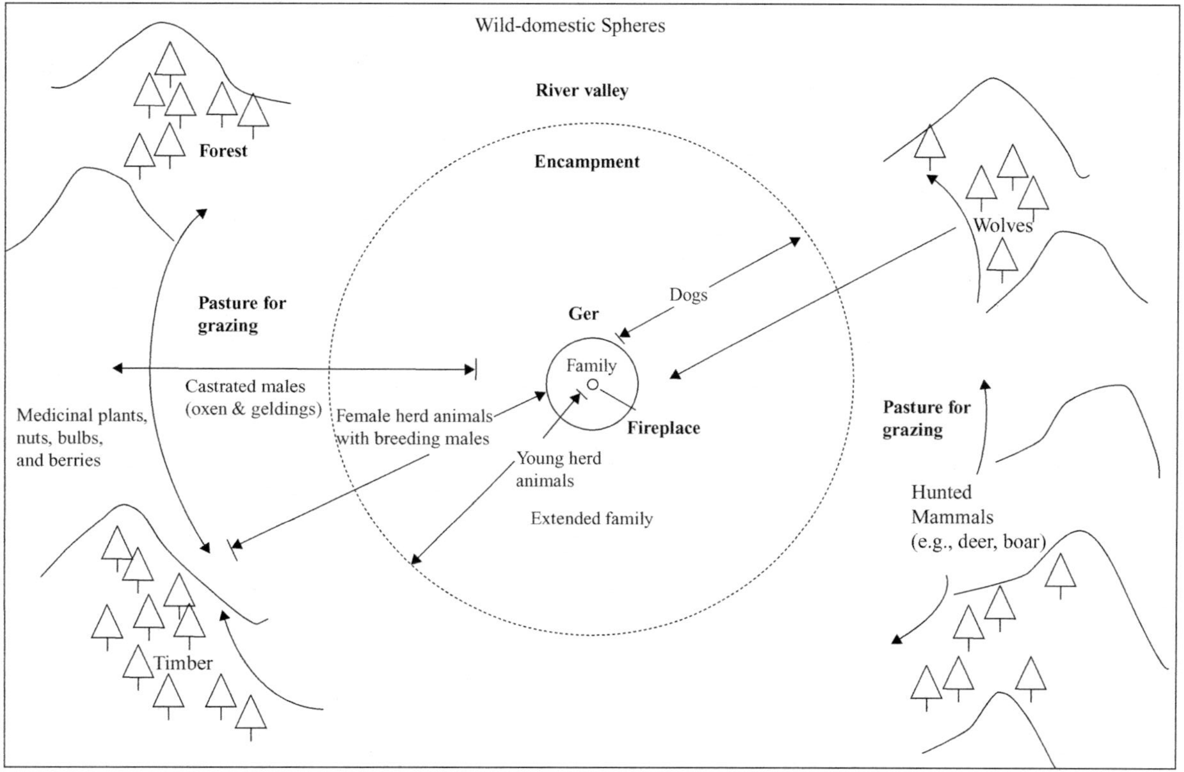

Figure 9.1. Conceptual diagram of wild–domestic spheres.

herder, Dalkhaa, told me a story of how wolves were captured and enclosed within a fenced area during the Soviet era to eventually use their pelts (essentially a wolf "farm"). The local herders were given money in exchange for wolf pups from dens and the wolves were then held inside this enclosure. Dalkhaa emphasised to me that these wolves in the enclosure were not tame; they kept howling, were afraid of people, and would always hide. It is clear here that Dalkhaa considered these wolves to be wild animals because they were afraid of people. To Mongolians this behavioural element of a fear of humans is a key factor as to whether an animal is wild.

ANIMALS AND PLANTS BEYOND THE DOMESTIC SPHERE

In the Khangai Mountains, herders supplement their pastoral diet by collecting plants and berries from the surrounding countryside over the short period of growth during the summer. Tending the sheep and goat herd gives an individual herder the opportunity to experience the surrounding landscape at a leisurely pace and to take note of the plants the animals forage upon. Herders in the Khangai make ample use of the "wild" plants in their valley, to treat the co-domestic humans and non-humans in a constant flow of multispecies interchange and interaction. They have a good knowledge of the potential uses of plants within the river valley that they migrate up and down from one year to the next. At the same time, the herd animals provide valuable nutrients and disperse the seeds of these plants in their dung, making the process ongoing, aligning with the Mongolian herder cosmology of a constant cycle of death and rebirth from one organism (or being) to another.

Herders gather nuts, bulbs, and berries that are farmed and cultivated as domesticates in other parts of Eurasia yet are still growing wild in Mongolia. Each herding family has specific locations where they know that they can find the bulbs or fruit that sprout up every summer. These plants provide necessary vitamins and seasonings that add variety to the herding diet. I would go with children to collect strawberries (*gÿzeelzgene*) from a nearby patch of forest. The strawberries were smaller than farmed ones but were just as tasty, perhaps even more so because we had gathered them ourselves. It took a lot of effort searching for tiny strawberry plants before one small jar was filled. The children also enjoyed digging for bulbs that were crunchy and resembled large spring onions but were commonly called potatoes (*töms*).[2] We scaled a small variety of pine to knock down little pinecones, breaking

[2] The official Mongolian name for this bulb is *Odoi caraana* and the scientific name is *Lilium pumilum*. The Mongolian strawberries are *Dornodin gÿzeelzgene* or, in Latin, *Fragaria orientalis* (Galbadrakh & Ganchimeg 2000).

the husk open with our front teeth to reveal a pine nut (*somar*) within. Just riding across the grassland revealed rhubarb (*balchirgana*), which was growing in healthy clumps. Watercress was gathered from the nearby river and mushrooms were collected from the hillsides in late summer and dried to give to cows when they were weak in the spring, or to provide strength after a herding woman has given birth. When I visited a neighbouring herding family, the whole family was covered with juice from the purple-blue berry (*ners*) they had been gathering, their faces smeared in the bright colour from eating whole handfuls of the berries. Instead of attempting to domesticate and alter these plants, they are left to propagate and spread, independent of drastic human intervention.

Mongolian herders view the world in relation to their extended family, which includes the herd animals. It is interesting to note that some of the common plants that herders encounter while out herding are named after features of the five kinds of co-domestic animals. The following are examples of common plant names referring to features of the herd animals: Lamb's Ear (*Khurgan Chikhi*); Lamb's Testes (*Khurgan Töms*); Billy Goat Beard (*Ukhan Sahal*); Goat Horn (*Yamaan Ever*); Cattle's Eye (*Ukhriin Nud*); Stallion (*Azargana*); 2-year-old Horse's Tail (*Daagan Suul*); Camel Knee (*Temen Obdug*); Camel Teat (*Temen Khükhü*); and Camel Tail (*Temen Suul*). Attributing such names to common plants gives a clear indication of the prominence of animals in a Mongolian herder's view of the world, particularly co-domestic herd animals.

Animals beyond the co-domestic sphere are only occasionally interacted with and this is usually for the purposes of hunting for food. Mongolian herders rarely eat birds or fish but they do hunt mammalian game. If deer or wolves are sighted while the herders are out herding, a few men will gather and ride out to hunt them, perhaps accompanied by a good hunting dog.[3] Young children actively engage in attempting to capture abundant ground squirrels (*zöröm*) by flushing them out of their burrows with water. Apart from hunting the frequently sighted wolf, herders occasionally hunt red deer (*ulaan buga*), musk deer (*khuder*), and wild boar (*zerleg gakhai*).

Most wild mammals remain deep within the surrounding mountains and forest, rarely venturing close to the encampments on the valley floor. In the Khangai Mountains of Mongolia, animals that are more commonly sighted are species of bird (*shuvuu*), ground squirrels (*zöröm*), deer (*buga*), and wolves (*chono*). Birds that scavenge from the carcass of a herd animal

[3] Not all herders I encountered hunt. A young herder, Nasaa, did not hunt and he expressed his regret that there were so few wild mammals within the valley (except wolves), which was at least partly the case because people from the city came to hunt for recreational purposes.

do routinely appear: kites (*eelee*); huge, lumbering vultures (*tas*); crows or ravens (*kheree*); and magpies (*shaazgai*). The wolf and the crow are perceived by Mongolian herders as having a mutual dependence upon one another. Dogsomjav told me that *kheree* often follow wolves and cooperate with them. Mongolian hunters say that the wolf eats prey with its eyes closed and only raises its head in order to look at the *kheree* that sits waiting on a nearby tree, as the *kheree* will warn the wolf of any danger. In return, after eating its fill, the wolf will leave the rest of the carcass to the *kheree*. *Shaazgai* sometimes call when they see wolves, alerting the herder to a wolf attack. Mongolians often told me that *shaazgai* had the ability to predict the future and bring good luck; for example, when a *shaazgai* sits upon a post and calls, the bird is predicting that the family will soon have visitors. Herders pick up knowledge of such birds because the scavenging birds feed from the carcasses of the herd animals.

As is the case for many people in Western cultures, some Mongolians are fearful of such animals as spiders, snakes, and mice. I gave a child a toy wooden snake that moved and was surprised that the women in the encampment were fearful of the toy snake and were tentative about touching the toy. Lhagva was afraid of many of the smaller animals, such as snakes, frogs, mice, spiders, and moths. She told me, "I don't like how the moths have big abdomens that are filled with eggs." She would ask me to get items from the storehouse so that she could avoid spiders, or in the night she would ask me to get rid of mice that were rustling amongst the cooking pots because she knew I did not have the same fear. Mongolians seem to be particularly afraid of mice and cats. Herding men who are not afraid of handling big oxen and horses many times their own strength can be afraid of mice. While the men from Choijo's encampment were scything long grass to make into hay, one of them came across a nest of tiny, naked rodents and let out a cry of surprise, leaping away from them. When I walked over and picked one up, cupping it in my hand, the men were amazed that I was not afraid to handle the "mice" (*khulgana*).[4] They explained their fear in saying that mice carry disease and get into peoples' food, yet this same fear is not applied to the marmot (*tarvaga*). Avoidance of mice may be a measure to avoid disease but, if so, it is ironic that it is the Mongolian marmot that harbours the enterobacteria that caused the bubonic plague (*Yersinia pestis*).[5] Like the wolf, mice and

[4] This aligns with Simpson's (1938) account that Mongolians colloquially refer to many species of rodents by the generic term of "mouse" (*khulgana*).

[5] Herders still hunt and eat marmot but they are careful to avoid eating the marmots' lymph glands. When hunting, they will not shoot marmots that do not give an alarm call, because the call is not given by individual animals that are infected and weakened by illness (Mayhew 2001). While I was in Choijo's encampment, a local doctor came and vaccinated the family members against the illness by means of an injection into their backs.

rats sometimes enter into the encampment, producing greater antagonism and fear than the marmot that remains within the proximity of its burrow beyond the encampment.

This fear is with regard to specific *kinds* of animal, as domestic cats are also animals that Mongolian herders fear. A few encampments do have cats, but the majority of herders do not have them. I was told a number of times by people that they thought a cat would scratch their faces or smother them while they slept during the night. I commented to Dalkhaa about having a cat in his *ger* and he replied, "The only advantage of a cat is that they catch the mice; the cat is viewed as an unlucky animal and bad for people." Tuvshin told me, "I don't feel any fear of camels, even though they are such big animals, but I don't like to touch cats ... there's just something about them." Mongolians do not have a fear of "otherness" that rests on a wild–domestic dichotomy, but they do have perceptions about the character and positive or negative traits of specific kinds of animal. In the following section I make a comparison between the Mongolian grey wolf (*Canis lupus*) and the Mongolian guard dog, involving a complex mix of both "good" and "bad" behavioural traits.

THE WOLF AND THE DOG

The wolf and the dog have engaged in a long association with humans. The dog's role is often one of companion, guardian, and protector, whereas the wolf is a competitor for game and a threat to livestock. The domestic dog evolved from the wolf and was the first animal to co-habit with humans, perhaps surviving by scavenging on refuse around human settlements (Coppinger & Schneider: 1995), or alternatively by adopting the role of guardian and watchdog in exchange for food scraps (Newby 1997). Genetic evidence has shown that there is a large degree of genetic variation in dogs in the region of East Asia (including China and Mongolia; see Savolainen et al. 2002). Just when, where, and how often initial coexistence between humans and dogs first took place is still under dispute.

The wolf has been a key subject for allegorical or metaphorical representation. An excerpt from *Of Wolves and Men* by Barry Lopez (1978: 210) nicely describes ancient Roman beliefs regarding wolves:

> If twilight is the time of the wolf, it is interesting to note that Mars, before he became the Roman God of War, was the agricultural God of the Lands that separated cultivated fields from wild woods. The twilight image of the wolf begins to resonate here with the sort of truth the mind both generates and delights in, because Mars' special animal was no other than the wolf.

To the Romans, the wolf represented the separation between the cultivated and the wild, the domestic animal and the beast. The planet Mars can be seen in the sky at twilight and this too is the time when people and livestock encounter the wolf.

In many cultures there are stories and legends of packs of wolves nurturing young "wild" children. This is an interesting reversal of the usual relationship of humans nurturing the canine. Here, the canine domesticates the human into canine society. One such legend exists amongst peoples of the Altaic language group (which includes Mongolians). In this legend, the wolf and the crow cooperate with one another to nurture a child; the crow flies over the baby with meat in its beak, followed by the appearance of a female wolf who lets the baby suckle milk. As Jila (2006) points out, this story is remarkably similar to the Roman story of Romulus and Remus, which features a baby suckling from a wolf and cared for by a woodpecker. The image of a female wolf nursing Rome's founding twins, Romulus and Remus, was famously inscribed on Roman coins. In Western tradition the wolf is represented as a link between the domestic and the wild, where wolves have the potential to become "domestic" and humans the potential to be "wild."

The Mongolian Wolf

Mongolians do not view wolves in a similar category to the dog, because they hold different status and roles in Mongolian herders' lives. The wolf is thought to be a "high" or "spirit" animal, whereas the dog is seen as an extension of the family. Herders recognise that wolves sometimes display good, and sometimes bad, characteristics. Using the word "wolf" could invite a wolf to attack the livestock. The common Mongolian word for wolf is "*chono*," but the wolf may also be referred to as the "sheep's assassin." Mongolians do not view wolves with fear and dread; their view is that wolves are afraid of humans, particularly humans with guns and "small red faces."

In legend, Mongolians as a people originally came from a wolf: This wolf was the Mongolian herder's first "father." In an ancient text from the mid-thirteenth century, *The Secret History of the Mongols* (Onon 2001), it states that the origin of Chinggis Khaan is from a bluish wolf, born through destiny from the heavens above. As Bira (2004) argues, this is no ordinary wolf but comes from the heavens and is thus a sky animal. This origin story makes the relationship between the herder and the wolf particularly complex and interchangeable. Herders kill wolves in order to protect their livestock, yet they originated from the wolf in the beginning. It is believed that both men and wolves are linked to the sky, or the "father" heavens (*Tengger*), whereas women and deer are linked to the "mother" earth (*Etugen* in shamanist tradition, or *Lus* in Lamaist tradition; see Even 1991). A

Mongolian friend, Tuvshin, used this origin myth as an example of how wild animals (and therefore nature) are an integral part of Mongolian culture and how nature is used as a tool for teaching future generations of Mongolians. The strength of a Mongolian wrestler is linked to that of the wolf. Wrestling incarnation rites from one generation to the next are concentrated into a legend that "after the strong wrestler has passed away, a wolf makes a nest in his chest and procreates there" (Tomikawa 2006: 105). The reincarnation of one wrestler to the next is precipitated by the spiritual powers of the wolf.

Prior to my fieldwork in Mongolia, I had the expectation that the wolf would have a presence in Mongolia similar to that in North America or Europe, where wolves were once a very real threat but as a result of their reduced numbers have become a background figure, mainly encountered through legends. During the Soviet era this was almost the case in Mongolia, as a national wolf hunt was organised every year (Pratt, MacMillan, & Gordon 2004). The wolf did become extinct in large parts of Mongolia, but since the end of the Soviet era wolves have rebounded in numbers (Hovens & Tungalaktuja 2005). The Mongolian government allows the hunting of wolves, and many Mongolians from Ulaanbaatar go on wolf hunting trips with high-powered weapons for sport. An image of wolves in Mongolia comes back to me with a nightmare-like quality. During a winter trip near Ulaanbaatar, my Mongolian companions and I pulled onto the side of the road for a short break. Looming out of the darkness was a macabre scene of twelve stuffed wolves, literally frozen in place with fierce, snarling expressions in the lights of our vehicle. No one from our vehicle said a word but quickly returned to the warmth of the Russian jeep and we drove away into the dark of the frozen grassland steppe in silence. The wolf hunter's intent was to display the wolves to passing traffic, so they might buy the wolf pelts.

In the countryside, however, herders do not usually kill wolves for sport; they kill them to prevent them from encroaching into the co-domestic sphere where they pose a threat to the herd animals. Mongolian herders recognise that wolves can be beneficial; herders say, for example, that wolves "tidy up" weak and diseased animals. On a number of occasions herders informed me that if the wolf disappeared, this would upset the balance and many plants and animals would disappear along with the wolf. In some areas, if herders find a den with wolf pups inside, they will shoot the pups but may leave one for the mother wolf to nurture (Kaczensky 2007).

Mongolian herders respect wolves for their intelligence and cunning. There is a Mongolian proverb, "A thief has free time but a wolf has the rain." In explanation of this phrase, the herders know that wolves have a keen awareness; they say that the wolf is more likely to attack an animal when it rains because a herder is less likely to be out watching over the flock. Herders have learnt the wolf's tactic of hunting in the rain and try to counter the wolf's tactic by venturing out in bad weather. It becomes a war between respected parties, of attack and counter-attack. A large amount of the herders' time and energy is spent guarding and protecting the herds against wolf predation. They call out while herding to deter wolves, and they avoid forest where wolves can hide in ambush. When wolves are heard howling in the night, the men stay up guarding the animals in their pens and howl back to the wolves, mimicking their ghostly calls to gauge the location of the wolf pack. The men yell out specific calls to calm the sheep and goat herd and to warn the wolves, particularly an elongated "*chaa*." Like farmers putting up scarecrows in a field, herders cut up plastic bottles into mini-windmills that move and spin on a pole near the pens to distract wolves that venture near the encampment. In Dogsomjav's encampment, a smelly mustelid's tail (*ömkhii khüren*) hangs from a pole to one side of the pen and is placed there to act as a deterrent against wolves, because the wolves can detect the pungent smell.

Dogsomjav described the habits of the wolf to me in the following way:

> The wolf is very smart. When a wolf attacks a pregnant sheep, the wolf may take the lamb from the sheep's stomach [and] then just eat the adult, allowing the lamb to grow up. I know of one herder, when he arrived at a sheep that was killed by wolves, there was a newborn lamb with no mother, so he took it back to feed it by hand. ... The wolves have a meeting time between December and February [the middle of winter] where they do not attack animals and just feed from the wind. In this meeting time of the wolves, a female leads the others. If a human kills the female wolf, then a male wolf will kill that human in retribution. During this meeting time, wolves will kill other bad wolves by fighting with each other. For the following few months the wolves just feed on insects and plants. The springtime is the main time when the wolves feed on newborn herd animals. (Field note excerpt: Dogsomjav's winter encampment, Dogsomjav, April 29, 2007)

Herders are particularly wary of wolves in the spring, as wolves often take lambs or kids when they have offspring of their own and are particularly hungry after winter. Herders informed me that they had also seen wolves take live kids and lambs back to their pups in the den to teach the pups how to handle prey. Wolves are frequently heard howling, sighted trotting around even in broad daylight, and attack the livestock in the dark of night. During my time at Dogsomjav's encampment, wolves attacked livestock on five occasions. At Choijo's encampment, Ganbar informed me that they had suffered twelve wolf attacks that year amongst the encampment's herds, including the death of six foals. Hovens and Tungalaktuja (2005) found that livestock are an important part of wolves' diets in Mongolia, particularly horses, because herders do not guard them as carefully as the more vulnerable sheep, goats, and cattle.

Mongolian herders often engage in conversation regarding wolves and the best way to treat wounds from claw marks and puncture wounds. The herders have a good knowledge of how to treat animal injuries incurred by wolf bite. A wolf attack is a major event. Dogsomjav is recognised by other herders as being particularly knowledgeable about medicinal and "magic" treatments for sick animals. The following is part of a conversation between Dogsomjav and a neighbouring herding woman, in which she gave some advice on the best ways to treat some cows after a wolf attack[6]:

Baasaa:	Yesterday a cow and heifer were clawed again. I was going to get tree mushroom [to treat the cows who had been previously attacked] and when I was on my way I could hear our cattle bellowing. . . . I spurred on my horse and raced over but the cow with her calf was already being clawed. A white creature was holding onto its back. It wouldn't get off even when I was shouting . . .
Dogsomjav:	If they're not afraid of yelling, perhaps only gunshots would work.
Baasaa:	That big white wolf was holding and holding and just wouldn't get off.
Dogsomjav:	Did you boil an ant's nest [as treatment]?
Baasaa:	Yeah, we boiled up everything together. I gave three litres of milk in the morning and evening.
Dogsomjav:	That's good, are they okay? [She motions towards the cows grazing in the distance behind Baasaa]
Baasaa:	Just their udders are torn. No canine tooth wounds, just marks.
Dogsomjav:	It will be fine if the wound is open and liquid runs from it.
Baasaa:	No liquid is coming out at the moment. The hindquarters of one cow were attacked and the wound is open. The others are so-so, but their udders have gashes.

(Dogsomjav's spring encampment, May 17, 2005)

[6] See A Day in the Life, available at www.cambridge.org/9781107000902.

Unfortunately, two of these cows died from their wounds, as bites from wolves can often turn septic from bacteria in the wolf's mouth. Herders leave a carcass that has been attacked by wolves; they informed me that, because wolves eat carrion, their mouths contain strong poisons that are dangerous to both humans and herd animals. The medicinal plants used to treat wolf attack are oriented towards combating infection and cleaning wounds. The use of "magic medicine" such as the ant's nest is thought to combat the effects of the powerful wolf. If the herder manages to kill the wolf that attacked a herd animal, the dead wolfs bile is smeared on the lips or tongue of the injured animal to assist in its recovery.

The cause of a wolf attack can be due to a herder not behaving in the correct manner. The wolfs presence challenges a family's strengths and weaknesses. As Divangar commented, "Nature sent the wolves to test the family. Is that herder strong, or lazy?" Here, "Nature" (*baigal*) holds the power, not the herder. The wolf is viewed by Mongolians as intelligent, strong, and courageous, as a "sky animal" aligned with *Tengger* (see Chapter 10 for more details on the concept of *Tengger*). If someone has a vision of a wolf in his or her dream, it is a prediction of good luck for the future. To gain a sighting of a wolf, even a dead one, is thought to bring the person luck. On one occasion, when a female wolf was killed by a neighbouring herder, Sanjaa, and brought back to the encampment, everyone crowded around the dead wolf, touching the wolfs coat and inspecting the sharp canine teeth, which is a way of conferring luck upon each person. The women of the encampment extracted the blood from the heart and stomach of the female wolf for medicinal purposes. The blood from the wolf's heart is used for prevention of heart attack and the tongue to cure toothache or an injured throat.

Sanjaa is an expert wolf hunter amongst surrounding encampments. He told me that some years he kills over ten wolves. He killed three wolves while I was living in the valley; ironically, during this time his encampment's herds suffered from a succession of wolf attacks. His strategy is to wait until after it has snowed, then to follow wolf tracks back to the wolf den to kill the adults and the wolf pups. If a herder kills a wolf, then that means the herder is lucky, because the wolf only allows itself to be killed by a person with equal luck, providing a balance for the death of the wolf. Sanjaa explained why he kills wolves: "The wolf is very lucky, so if I kill a wolf then my life will also be very lucky." When I asked why the wolf is thought to be lucky, he explained that it is because it remains part of the mountains and forest. He added, "The wolf is afraid of the human face, so only appears for people who are lucky. The wolf thinks 'I am afraid of small, red faces but the human is not strong like me, the human is like a two-year-old steer" (Sanjaa, spring encampment, April 30, 2007).

The Mongolian Dog

The breed of dog in Mongolia is thick set, with a thick winter coat and curled tail like a husky. The stature and colouring is often similar to that of a Rottweiler. Many Mongolian dogs retain "second eyes," or two lighter coloured patches above the eyes, which are thought to be favourable for protection against wolves (Boyd & Houpt 1994). Dogs in Mongolia do not guard the herds as they graze: In fact, the dogs have little engagement with the herd animals. In other parts of the world, dog breeds are specifically intended to guard flocks against wolves; one such example is the large, white guardian dogs from the Pyrenees who grow up and mingle with a sheep flock, forming a strong interspecies bond with sheep (Coppinger & Schneider 1995; Haraway 2003).

Naraa described the difference between dogs and wolves to me in the following way: "The dog is an honest friend of people, is smart and guards the family but the wolf finds its own food, so [it] is strong and lucky." Herders prefer not to eat either wolf or dog meat, but specific parts of the body are extracted as medicine and the pelt may be used for hats and clothing. The medicinal properties contained within the internal organs of a wolf are more powerful and provide better results than the same internal organs from a dog. Dogsomjav explained to me that the wolf is not fed by humans and has to struggle to find food for itself, making the wolf "strong." Dogs do have spiritual meaning for the Mongolian herder, however, particularly when a good dog dies. Upon the death of the dog its tail is cut off and laid under its head in the hope that the dog will become human in its next life, in acknowledgment of the fact that a distinctive feature of the human form is the absence of a tail.

Relationships between Mongolian herders and dogs are based on the dog's relationship with a specific family. There are often one or two dogs to protect each family's *ger*, and there are often between two and six dogs within each encampment. As the primary role of dogs in Mongolia is to guard the encampment against human strangers and wolves, Mongolians are wary when visiting any neighbouring encampment and will refrain from dismounting from their horse until they have yelled "hold the dogs!" (*nokhoi khorio*) and someone has restrained them. A valued dog is one that barks at strangers or wolves in the vicinity of the encampment. The Mongolian dog guards against wolves but it too can become prey to wolves. The dogs are usually untied in the evening and roam around the encampment during the night. If wolves howl nearby, this may set off the dogs in nearby encampments with rounds of barking throughout the night. Dogs are not commanded to refrain from barking during the night because their activity can warn the family of any danger to themselves or their herd animals.

The wolf often features in herders' conversations, holding an almost mythical status in their lives, yet is still a very real presence for herders. It is the Mongolian dog who protects the extended family encampment, including both herders and herd animals, from the wolf's encroachment into this co-domestic sphere. The wolf and the dog occupy different roles in the herders' lives, one as sacred but also combatant and competitor for livestock and the other as guardian and protector. The difference between domestic and wild in the Mongolian landscape is less a physical or conceptual barrier and more a behavioural difference in the animal's relations with their human counterparts.

TREATMENT OF ILLNESSES AND AILMENTS

To Mongolian herders, the force of nature can place judgement upon whether a person is behaving in a correct or incorrect manner and has the power to inflict punishment upon a person who does not abide by rules and traditions. Powerful forces have the capability of inflicting illness or death if any person does not behave according to specific practice and with due respect towards Nature (*baigal*). I developed a cold and fever soon after arriving at Dogsomjav's encampment in spring. I had been out herding with Dogsomjav, beyond the secure surrounds of the co-domestic sphere of the encampment, on one of my first herding forays. I reclined on my back, while Dogsomjav sat with legs stretched out in front of her as we whiled away the time with the sheep and goats grazing nearby. I became suddenly ill after we had ridden back to the encampment. I thought it was probably due to the new environment and the extreme temperature changes of the Mongolian spring. When Dogsomjav saw that I was ill, she told me gently that herders do not lie on the ground because there are bad "forces" carried down from the mountains and forest in the spring time and lying down can make one susceptible to them. She gave me a strong, hot tea made from cranberry leaves (*ansnii navch, Vaccinium vitis-idaea*). After I gargled the brew I felt revitalised and recovered from my illness within a couple of days. As this example demonstrates, herding families in the Khangai Mountains are capable of treating illnesses, and they make use of the powerful medicinal plants from the surrounding mountainsides. According to Humphrey with Onon (1996), the view from shamanist tradition is that human or animal illness is intended by the mountain (or mountain spirit). Medicinal berries, such as cranberry, that are found on the slope of a mountain are "given" by the mountain and a deer is "sent" for a hunter to find it.

Care of the herd animals includes the provision of sound shelter for the animals during the winter; Western medicines (such as penicillin); treating illnesses with locally collected medicinal herbs; and through ritualised "magic medicine" (*em dom*) practices. Treatment often involves a combination of all of these techniques. In subsequent text I illustrate these techniques used by the herders in more detail, with particular emphasis on treatment by use of medicinal herbs collected from the surrounding countryside and magic medicine practices. Choijo's extended family has a large number of animals. Daily the men and women put time into treating the herd animals for various illnesses, such as problems with the eyes, diarrhoea, worms, gouges from sticks, rubbing from the tethering line (*zel*), or wounds caused by defensive kicks from other animals within the herd.

Herding families have access to local veterinarians but their role is minor in comparison to the treatment of animal illnesses by the herders themselves. When I asked Lhagva how she attained knowledge of the animal's illnesses, she replied, "It's not difficult to gain medicinal knowledge because we look after our animals all the time. We have to know the animals' illnesses and how to treat them when they look particularly tired. If our animals are dangerously sick, then we go to see the veterinarian" (Lhagva, autumn encampment, October 21, 2005). In Bulgan, Aagii's father would visit once every month or two during the summer season but less regularly in winter. Usually this was to administer general worming or vaccination treatments for all the encampments within the valley. He would examine the state of an animal if it was sick at the time of his visit, usually giving verbal advice, as the herders were generally not able to pay him for expensive Western medicines. Unlike the Bulgan field site, where Aagii's father was the veterinarian, in Arkhangai I did not encounter the local veterinarian visiting Choijo's encampment. Naraa said the vet wasn't much help with their animals, which meant that they learnt to treat the animals themselves.

Herders combine their knowledge of both the animal's physiology and its behaviour to aid in the animal's survival and well-being; for instance, the health, strength, and speed of a horse are related to the health of the lungs, heart, liver, spleen, and kidneys. According to Bayarsaikhan (2006), herders examine the size of the nostrils to check the health of the lungs. If a horse's eyes are dark-brown or reddish-black, then it has keen eyesight. A calm horse that is rarely startled has a large heart. If a horse's tongue is white, then the liver is in good condition. If a horse has thin, hard ears then the horse has a small kidney. Herders take the sheep and goat herd slowly in autumn and avoid sudden stimuli because if they are moved around too quickly they will not retain their weight over the long winter. Lhagva told me that when they process and cook an animal in autumn, if fat from the sheep congeals on her hands it means that the sheep may be losing too much fat, so they know to move the animals at a slower pace while out herding.

Bloodletting is often used on horses when they become lame or have developed a cold. A horse's lameness is treated by extracting blood from the frog of the horse's hoof. Herders informed me that if a horse stops too suddenly after galloping without being walked around, or if the saddle is taken off too quickly, the blood goes down to the horse's feet. Symptoms include a horse repeatedly raising its hooves in turn while stationary, lying down and getting up repeatedly, or limping while being ridden. I witnessed four horses tied to a hitching post that had small incisions made with a knife at key points on their bodies, the blood spurting in an arc onto pools of blood that had already accumulated on the ground. The incisions were made on the forehead, upper lip, chest, and upper legs to let out "bad" blood. Similarly, the Nuer of East Africa treat sick cows by drawing out the blood: "Bleeding is designed to cure a cow of any unfitness by letting out the bad blood of the sickness. Also, Nuer say, it makes the cow fat, for the next day it will be more lively and graze avidly" (Evans-Pritchard 1940: 28). When I asked Aagii, a veterinarian from the Agricultural University, about the effectiveness of bloodletting, he thought that it was effective, "particularly in the treatment of lame horses. In fact many Mongolian veterinarians use bloodletting as a technique."

Treatment of illness is often dependent upon the circumstance, such as the degree of illness, the species concerned, and the practical knowledge of each individual herder. Between the two field locations of Dogsomjav's and Choijo's extended family encampments, each employ a variety of remedies for treating infected eyes. In Dogsomjav's encampment they spit salted water into the cows' or calves' eyes and if this does not work they resort to scarce and expensive penicillin. In Choijo's encampment, after the morning's milking the women treat the calves' infected eyes by tilting each calf's head to one side, then sprinkling sugar onto the eye. If the infection continues they rub charcoal or pour milk onto the weeping eye. Similarly, herding mothers regularly treat eye infections in their babies by squirting breast milk into their eyes. In one instance, Davaa collected her baby's urine and poured it into her baby's eyes to stop them from weeping. When I asked Davaa why she used urine as a treatment, she explained to me that Mongolians traditionally drank their own urine to prevent illness, and some herders still do so.

Mongolian herders are also adept at treating injuries. Herders use heat and minerals to stop the blood flow, to kill any bacteria, and reduce infection. Larger animals, like horses and oxen, tend to be treated by the men, as a number of men work together to catch the sick animal, hobble it, and then push it over onto its side and hold it down on the ground, while one of them applies the treatment. The following treatment was administered to an ox that had slipped in the mud and snapped off its horn while hitched to a train of oxen. The horn was huge (greater than 30 centimetres long). Bone had snapped off with the horn and the ox was bleeding profusely from the remaining stump:

> Because it was Sumya's ox he washed the horn and the ox's face that was covered in blood with "yellow milk" [*sha suu,* or *aarts,* the run-off from curd], [and] then packed the wound with ash from the fire to stop the flow of blood. A black mixture [which looked like iodine] was smeared around the base of the injured stump [by] using fleece wrapped around a stick. The stump was then covered with felt and Amaraa quickly sewed a protective cover from the sleeve of some clothing and it was placed over the severed horn. This was then kept in place with twine. I doubt that the local veterinarian could have done a better emergency repair job! (Field note excerpt: Choijo's summer encampment, July 24, 2005.

MEDICINAL PLANTS AND MAGIC MEDICINE

The Khangai Mountains in Mongolia have an abundance of medicinal plants and herbs that the herders make use of in their daily animal husbandry practice (see Table A.2 in the Appendix). There are over 600 species of flora with recognised medicinal properties unique to Mongolia (Ligaa & Tsembal 2003). Medicinal plants (*emeenurgamal*) are often collected by herders while out herding. The appropriate time for collecting is before the plant's stamens collapse at the end of summer (around August); otherwise, the potency of the herbs can decline. Medicines are prepared by drying and often boiling the plants to treat illnesses of members of the extended family. Particular plants and herbs impart specific cooling or heating properties; for example, in the case of fever, substances that are cool and dry are required. A feature of this medicinal care is that many of the herbs, berries, roots, and flowers are used to treat the illnesses of both the human family and the five species of herd animal.

There is not a strict line between care provided for humans and care provided for ungulates, yet one plant may be used to treat a particular type of illness in an ungulate species and used for family members to treat a different type of condition. This is due more to a disparity in physiological make-up in terms of differing digestive tracts and metabolism in ungulates compared with humans, than to a difference in the degree of importance placed on the being itself. An example of slightly different treatments of human than of ungulate is in the use of the plant *Tsagaan tseene* (*Paeonia lactiflora*), which is used to treat liver or kidney problems, specifically when a person is having trouble urinating; it is also given to herd animals to give them energy in the spring after the long, hard, winter months.

There are many treatments that help herd animals to survive through the spring, once all their reserves have been depleted during the extreme conditions of winter. Herders prepare medicinal plants and valuable nutritional plants (such as the nettle, *khalgai*), mixing them up with bran (*kheveg*), or they make a strong soup, such as horsemeat soup. A common method is to use different kinds of onion (such as *mangir*, or mountain *songino*) that can be collected from the surrounding hillsides, which is then mixed with dried sour curds (*aruul*). The run-off from sour curds (*aarts*) is used to cleanse wounds, because the bacteria in the curd can fight infection. Mongolian vodka (*nermel*) is often used in combination with other measures to treat diarrhoea and stomach-related illnesses in both family members and herd animals, because the vodka kills unwanted bacteria in the stomach.

So-called magic medicine and the heightened power from medicinal plants assist in counteracting the negative forces attached to an illness. Tuvshin, a lecturer at the Agricultural University, commented on these magic medicine rituals, saying that "herders think 'this animal will be strong' and this prepares the herder spiritually ... if they think that they did wrong in the past, then [an animal's illness] may be the consequence, so they try spiritual treatments." The following field note extract illustrates a mixture of medicinal techniques and the application of these methods in treating a particular herd animal[7]:

> Today Lhagva treated a goat that was ill from an inflamed udder. The goat couldn't walk properly from the pain in her udder and Lhagva recognised that the nanny had a temperature. Lhagva wanted to give the goat a penicillin injection but she had none, so she force fed her some crushed paracetamol on a spoon. She then stroked the nanny's udder with two dried bear (*Ursus arctos*) paws and recited an incantation of "*em dom, em dom*" ("magic medicine," repeated). Incense was then rubbed onto the udder from an incense burner with a Tibetan inscription on the base. She then milked some of the milk from the udder, which looked curdled and frothy. I asked where Lhagva had learnt this treatment and she said "from both of my parents." She explained that the cause of the "dirty" udder may have been from someone hitting the nanny goat with a whip, as the udder is known to be a very sensitive part of the body. (Field note excerpt: Dogsomjav's spring encampment, May 8, 2005)

Similar methods of magic medicine are used for female herd animals and humans when they have problems with milk production. The treatment is a

[7] See Medicinal Treatment, available at www.cambridge.org/9781107000902.

variation on a common cause, a "dirty" udder or breast. A dirty udder can be due to any number of reasons, one of which was given by Lhagva: The nanny was hit by someone with a herding whip (*shelbuur*). In this case the negative action has to be dealt with by a counteraction, through "magic" treatment with bear paws and incense. When a cow has sore udders she is stroked on the udder with the top of the foot. Shoes and socks are removed and the sweat from the foot is wiped on the udder to "cleanse" it (Mongolians believe sweat is a powerful substance and wipe the sweat from winning racehorses to bring themselves good luck; see Chapter 7). Women who are pregnant, or have problems giving milk to their babies, are also treated with bear paws. According to Lhagva, twin babies' socks are an effective means of cleansing a breast or udder, as newborn twins are particularly auspicious, signifying fecundity and prosperity. Herders steal the socks from the twins' parents, as the treatment is not effective if the parents are asked for permission prior to the use of the socks. The family is prepared to have their babies' socks stolen and so leave socks strategically lying about to be easily taken without embarrassment.

Before my fieldwork I had not predicted the importance of care and protection in the daily lives of Mongolian herders and their livestock. A major part of the herders' lives is engaged in the treatment of animal illnesses, using a vast array of techniques and an impressive knowledge of local medicinal plants. This knowledge is passed down from one generation to another and from one herder to another. Aagii emphasised this to me:

> Traditionally, Mongolian methods for preparing young people to become good family members involve traditional education and this is how they learn to treat animal injuries. They maintain [this knowledge] from father to child and from these children to their children ... In recent years Mongolian veterinary services have developed, so we have some clinical services but everyday life is very different. Herders often have to just manage themselves. (Aagii, Ulaanbaatar, November 2, 2005)

A fundamental part of the co-domestic relationship is in nurturing health and general well-being. If the herders treat other beings and the landscape around them with respect, then herders believe that their extended family will remain strong and healthy.

CONCLUSION

To Mongolians, Nature (*baigal*) holds the power, not humans. Herders cannot attempt to harness or conquer nature, because it is a part of the existence of all beings. The domestic is distinct from the wild but only in degrees, and this degree is often gauged in terms of the extent of an animal (or plant's) association with humans. To be a co-domestic animal is to engage with humans and to be encompassed within the landscape, or co-domestic sphere, of the encampment (*khot ail*). Beyond the encampment, animals are perceived as being afraid of humans, but in their struggle for survival these animals have their own power and strength. A good example of this relationship between domestic and wild is the comparison between the dog and the wolf. The dog is perceived as an honest protector of the co-domestic sphere of the encampment, whereas the wolf is perceived as cunning and strong but at the same time afraid of humans, particularly their faces. This does not mean, however, that the landscape beyond the encampment is not engaged with, as herders in the Khangai Mountains hunt animals and collect bulbs, nuts, berries, and medicinal plants. These foods are not just viewed as a resource but are an offering from powerful spirits, such as the mountains, forest, or the weather. This power is recognised in the ability of the medicinal plants and animals to assist in the illnesses of not just the herding family but the herd animals as well.

10

THE SACRED ANIMAL

INTRODUCTION

A pictorial depiction of *Maliin Banzuragch*, a Mongolian Buddhist god and protector of the five kinds of co-domestic animal (*tavan khoshuu mal*), is situated next to the shrine at the rear of Dogsomjav's *ger* (see Figure 10.1). The top centre of the picture depicts a god riding a horse-like beast into the heavens. Key elements of fire and water are central components of the drawing. Each of the five kinds of herd animal flanks the central gods: a female horse, cow, sheep, goat, and camel, each nursing her own young. Below, and to one side, a figure is revealed stealing livestock and on the opposite side a wolf has been slain by arrows. This cloth *thangkar* helps protect Dogsomjav's precious herd animals from harm.

The practices I detail within this chapter offer a means of understanding the underlying attitudes and relationships that Mongolian herders have with the animals they herd. I use the term "spirituality" with caution, as I intend to convey the concept of beings containing an "indeterminate energy" (Humphrey with Onon, 1996) or heightened power. A crucial thought that nagged me throughout my fieldwork was this: How can these herders kill and eat the animals that they have known since birth, to whom they have given individual names, and who are referred to as individual personalities with specific characteristics? In seeking to answer this, I examine the ritual surrounding the death of an animal and, in particular, I focus on the only animal within a herd that is singled out for special treatment and not killed, the sacred (or *seter*) animal. This animal is easily distinguished as being different from the rest of the herd because it has a blue sash (*khadag*) around its neck.

Figure 10.1. Photograph of a pictorial depiction of *Maliin Banzuragch*.

THE MONGOLIAN CALENDAR

Mongolian herders do not tend to describe themselves as being religious. This is partly because they were not permitted to subscribe to a religion during the Soviet era, in accordance with the government's communist ideals (Becker 1992). In daily life, however, many actions and events are related to belief in phenomena beyond the physical being. Every *ger* that I entered had a Buddhist shrine at the rear of the *ger*, accompanied by pictures of Mongolian Buddhist gods, with small offerings and butter lamps at the base of the shrine. I asked Naraa about whether her parents had sacred (*seter*) animals; she replied, "No, they didn't have *seter* because it was the time of communism and [as she motions under the bed] we had to hide our statues and shrines behind and under things" (Naraa, summer encampment, July 27, 2005). A number of times Naraa invited a Buddhist monk and an elderly herder who could read Tibetan sutra to bless the family, the herd, and the nearby sacred tree. Naraa's beliefs fuel the traditions of the rest of the extended family, so that all participate in blessing ceremonies. Members of Dogsomjav's extended family have a vast knowledge of wild medicinal plants and so-called magic medicine (*em dom*) remedies. Dogsomjav would circle her grandchild above the fire and burn mercury in a spoon while chanting to prevent the baby from crying and being afraid. When I enquired about Dogsomjav's family's beliefs, Davaa said her mother was not particularly religious at all and Lhagva, when pressed, said she herself was Buddhist.

The main element of the herders' lives that is a concrete marker of religious practice is the Mongolian calendar. This is a pocket-sized booklet and is often wedged between two poles in the ceiling of the *ger*. The main calendar that herders refer to is *Mongol Zurkhain Tsag Toonii Bichig* (Esteemed Writing of the Mongolian Astrological Calendar), written by a venerable Mongolian Buddhist monk who specialises in astrology (Tervish 2006). The herders did not profess to have a great deal of knowledge about spiritual realms and would answer my questioning with "a lama [or Buddhist monk] would know," or "it depends upon the family." The calendar is significant because it provides a link between the teachings of Buddhist monks and the practical application in the herders' everyday lives and their dealings with the herd animals. Herders do not refer to the calendar daily but only when required, and they take account only of the aspects that are relevant to their practical existence in the world. The details that follow here are gained from the knowledge of herders regarding the calendar, not from Buddhist monks who study the astrological details in greater depth (refer to Chapter 1 for Mongolian links with Tibetan Buddhism).

The animals represented in the Mongolian calendar are the mouse (*khulgana*), cow (ox; *ükher*), tiger or snow leopard (*bar*), hare (*tuulai*), dragon (*luu*), snake (*mogoi*), horse (*mori*), sheep (*khoni*), monkey (*michi*), chicken (cock; *takhia*), dog (*nokhoi*), and pig (*gakhai*). Each of the animals has its own energy, personality traits, gender, and compass direction. There is a well-known folk tale in Mongolia called "Why the Camel Rolls in the Ashes" (Altangerel 2006) that gives an explanation as to why the camel does not feature as one of the twelve animals. The camel is said to have been given a combination of the attributes from each one of the twelve animals: the ears of the mouse, the stomach of the cow, the paws of the tiger, the nose of the hare, the body of the dragon, the eyes of the snake, the mane of the horse, the wool of the sheep, the hump of the ape, the head-crest of the cock, the crooked back legs of the dog, and the tail of the pig. These twelve animals are an interesting selection of wild, domestic, feared, and revered creatures.

According to the solar–lunar calendar, these symbolic animals are divided into twelve years, twelve days, and even twelve hours within a day, forming a layered, grid-based configuration. The years, days, and hours are combined with five elements (*makhbod*), and each one of these is aligned with its own colour and compass direction. The animals are usually referred to by an element in a "male" year and a colour in a "female" year, which alternates from year to year, for instance "fire dog" or "red cow." Here the importance of colour is evident in how the colours relate to representative animals. This identification on the basis of colour aligns with the importance of colour identification for herding purposes (see Chapter 4).

Neglecting to consult the calendar may result in conducting an activity on a "bad" day. Carrying out an activity on an inauspicious day could cause the animals to become ill, extreme weather conditions, or the livestock to be attacked by wolves. Logically, conducting a horse-related activity on a horse day will be good for horses. Even particular hours are "good" or "bad" hours in which to conduct activities. A bad hour, for example, is one that is designated as a "black dog hour." Conducting activities according to an auspicious day, or hour, as well as being mindful of correct procedures, means that the entire herd and extended family can maintain a prosperous and balanced existence. Auspicious dates are detailed in the Mongolian calendar for castrating animals, cutting people's hair, clipping the wool from animals, making the move from one encampment to another, slaughtering animals, or cutting hay.

DEATH OF AN ANIMAL

Herders believe that a strong, healthy herd is a reflection of a herder's own behaviour and personal character. To gain an idea about the numbers of herd animals that are killed or lost from the herd during the year, I asked the herding families how many animals had been slaughtered or exchanged, died from illness, or had been eaten by wolves in the past year. I expected this task to be straightforward but found that herders were particularly uncomfortable about answering such questions. After questioning the first few families, I was reluctant to continue with my probing because it was clear that the herders did not want me to ask such questions. I gauged that they were not being exacting about their answers, underestimating numbers, or answering that there were no animals at all within a category. It was obvious to me that their answers were inaccurate. I knew quite a few animals had died from wolf attack because I had seen the carcasses while out herding, and I had witnessed the deaths of many of the newborn animals and mothers from illness in spring. The herders' unwillingness to answer, or to answer by saying that there were virtually no losses, was indicative of their wider discomfort regarding the deaths of herd animals. From the herder's perspective, a true account of the number of deaths would have been a reflection of their failure to look after the animals well enough. Because every life within the herd matters, herders are unwilling to admit that any of the animals have died under their responsibility. Speaking too openly regarding the deaths of herd animals would be equivalent to breaking the trust of the herd animals. If an animal dies from illness or is attacked by wolves, this is sometimes interpreted as a bad omen by the herder, perhaps caused by some negligence or inappropriate behaviour by the herder.

Adherence to skinning and butchering an animal appropriately, sharing the meat with other members of the group, and consuming the animal entirely without wasting any edible or usable parts are features of animist societies (Ingold 1994; Nelson 1983; Serpell 2005). All of these practices are in accordance with Mongolian customary traditions to ensure that the animal is treated with respect and that there will not be negative consequences resulting from inappropriate treatment of the animal in death. I was graciously excused for my initial lack of knowledge of Mongolian practices and ways of conducting oneself, but herders were careful to inform me of the correct procedures when it came to injury, castration, or slaughter of herd animals. Choijo requested that I did not film or take photographs on a few key occasions: the treatment of an ox that was bleeding profusely after its horn had snapped off; when any of the animals were being slaughtered; and during the castration of the horses.[1] Naraa explained that a previous visitor to the encampment had taken photos of the horse castration ceremony and some of their horses had died soon afterwards. A Buddhist monk had advised her that the horses had died because photos were taken during the castration. Naraa described Mongolian herd animals as being "very sensitive," implying that they had to be treated with particular care. Kaczensky (2007: 327) also notes that, during her time researching in Mongolia, "it became quite clear that people typically do not like to kill animals or to be associated with killing animals. Taking pictures of a person slaughtering livestock or of people selling meat is not much appreciated."

I asked Chinzorig, a local Buddhist monk, what form Mongolian people generally want to reincarnate into in their next life. He answered that it would be best to come back again as a human but otherwise as one of the "five kinds of animal" (horse, cattle, camel, sheep, or goat) because these are the

[1] I was careful to take video footage or photographs regarding aspects of the herding families' daily lives only in those instances in which the family felt comfortable. I refrained from filming any events that the herders felt sensitive about, out of respect for the families' beliefs and the herd animals themselves.

closest animals to people and act as protectors to the spirit world. Herders did not feel qualified to answer such questions and felt uncomfortable doing so. When I asked herders what happened to the animals after they died, I was met with silence; it would have been too awkward to pursue my questioning any further. Amaraa said, partially admonishing me, "we never speak about what happens to an animal after it dies ... just as we never speak about an animal personally before it is killed for food."

Retaining respect for the animal is important in death. To feel remorse, or sorrow, for the animal's death is to not respect the animal in its life. Once the animal is selected out from the rest of the herd to be killed, it is not personally referred to by name. To ensure the correct procedure is maintained, herders kill an animal according to customary traditions. Any Tuesday is a black day for the family, so animals are not slaughtered on this day. Children are not to engage with an animal that has been selected for slaughter. Davaa told me, "When I was a child, my father told me not to look at the eyes of the animal who will be slaughtered, otherwise it's a bad omen."

Only men are permitted to be present for the actual slaughter of an animal.[2] The animal is taken to one side of the front door, to the south-east side of the *ger*. To slaughter an animal, one man holds the animal on its back with the front legs outstretched. Another herder makes a small incision with a knife near the sternum and forces his arm through the diaphragm to snap the aorta that leads from the heart. Larger animals, such as horses or cattle, are knocked on the head with a hammer to initially stun the animal. Dalkhaa told me that he knew of the technique of killing an animal by slitting its throat, as this was the method used by Kazakh Mongols. He explained that the technique of an incision in the stomach resulted in a quicker death. Spilling blood is also insensitive to the animal and wasteful, so the slaughter technique of herders in the Khangai is to ensure that a minimal amount of blood is spilt. A remarkably similar attitude toward killing reindeer was related by Alexander King (2003: 140) in his ethnography on reindeer herders from Kamchatka, Siberia:

> I thought about the scene I had just witnessed in contrast to the slaughtering of cows at my parents' small farm [in the US]. It was not less gory, but it seemed less gruesome. Instead of being a cold, mechanical routine, I saw the deer slaughtered with love and respect. Every deer is connected to an individual owner, and the physical consumption of reindeer flesh by humans is organized by the spiritual connection between deer and humans.

2 Generally, the herding men would slaughter an animal when I was out of sight and engaged in other tasks. They let me observe the slaughter only when I requested that I do so for research purposes.

In Mongolia, it is the men's task to remove the outer hide with the fleece from the animal and ladle out the blood from the inside with a small bowl. The women then take over the task to clean the stomach and intestinal contents and prepare the innards for cooking. Children play nearby while their fathers and mothers work on processing the meat, which means the children learn the cycle of life and death from a young age. A small portion of what is considered to be the best part of the animal, such as a piece of fat, is placed before the Buddhist shrine inside the *ger* and a butter lamp (*zul*) is lit in honour of the animal. The women told me that the light from the butter lamp assists the spirit of the animal to find the way to reincarnate into another being. Graham Harvey, in reference to animist practice upon the death of an animal, comments that "culturally appropriate forms of respect are offered, and further respectful acts promised at and after death. All this is predicated on the ability of humans and animals to communicate, on the reciprocity of their relationships, and the mutual intelligibility of their notions and performance of respect" (Harvey 2006a: 147).

The animal is valued in that it is killed according to customary tradition, through not speaking about the death of the animal but also by not wasting any *parts* of the animal. Only the bones and the digesta inside the stomach and intestines are not eaten. The blood and innards are cooked up and distributed amongst the extended family. Portions are taken to members of any neighbouring herd encampments that are nearby (and who are frequently related consanguineally or through marriage). This sharing of the animal is an important dimension of respect for the animal, in that none of the parts are to be wasted.

Unlike Western societies that consume young animals in the form of lamb or veal, Mongolians do not eat animals that are under one year of age. When I told Tsetsegmaa that people in Western countries eat lambs, Tsetsegmaa said quietly, with tears welling up in her eyes, "we love our young animals, so we couldn't eat them." She must have thought it a strange practice, as she was being so careful to nurture some weak lambs that were sleeping beside the warmth of the hearth in front of her. It would be counter-intuitive for a herder to kill them and eat them before they had produced any young of their own, when the animals have not yet lived a full life. If a young lamb dies from weakness or illness, the herder then utilises the hide but does not eat the meat. The hide from the dead lamb is immersed in curd to retain its softness and after some time the hide is kneaded to make it pliable for sewing the soft wool into winter clothing. Parts of the animals have endless everyday uses: cow hide is stretched into leather strips to make a horse's bridle or a lasso (*buguil*); plaited horsehair is used to tie the wooden lattice inside a *ger* together; knuckle bones are used as a tie in the ornamental currycomb; sheep wool is beaten and pressed into felt for the lining of a *ger* or for warmth inside boots; an ox's horn is used as a bottle for feeding orphaned young animals.

Even though herders look after an animal from birth and give the individual a specific name while the animal is *alive*, a dead animal's carcass is treated as food, or a product to be used in daily life (see online video segment 6. Lhagva the Herder).[3] In life the animal is a sentient being that can communicate with a herder, express emotion, and exhibit character or personality. Upon death the animal's body becomes food, something that is necessary for subsistence. The animal's death is an important contribution to the sustenance of life. When an animal is slaughtered (*mal nyadalgaa*)[4] in Mongolia, every part of the animal is utilised, even the head.

> None of the family are squeamish about eating the head of an animal, even though the individual can be readily identified, based on the coat colour and markings of its head. A recently slaughtered animal was left with its severed head and legs from below the knee lying near the fire until the women had time to prepare it for cooking. I couldn't look at the disembodied head when I sat in the *ger* and studiously looked elsewhere. The head in particular had an impact on me, as I could recognise the animal who had once belonged to the severed head. In preparation for eating, a goat's horns are first removed and then the head and legs are put in the hot embers of a fire. The head is removed from the embers periodically and hair seared from the skin with heated metal. The head and legs are then boiled. The fat, skin, and brain on the head are particularly tender parts of the animal, so the family crowd around to take turns in eating the brains with a spoon, while holding the head in one hand, or pick off the tender flesh from the goat's cheek with a paring knife. (Field note excerpt: Dogsomjav's summer encampment, May 5, 2005)

Once dead, the animal is no longer sentient; it is made up of mere bodily parts to be consumed: meat, blood, and innards. However, the animal's essence, spirit, or energy is retained within the animal's bones. According to Even (1991), the bones are where the "soul" is situated (I prefer the terms "spirit" or "energy" rather than "soul"; see Chapter 1 regarding animism). He states, "The soul, which is believed to have its seat in the bones, is thought to go on living after death, which is only the loss of the envelope, i.e. the

[3] See Lhagva the Herder, available at www.cambridge.org/9781107000902.

[4] There are many Mongolian words for slaughtering an animal, such as *mal gargakh, am'tan nyadlakh, mal tukhuurugch, alan khyadakh, idesh khiikh*, or *khanakh*. The word "*chimkhekh*" is used, which literally means to pinch, to seize, or to tweak. This is similar to the English euphemisms for death, such as "loss," "demise," "pass away," and "decease."

flesh . . . the integrity of the bones is a basic condition for the proper preservation of the soul and its eventual reincarnation (Even 1991: 185). Here, it is interesting to note that the bones are the only part of the body that a Mongolian herder cannot eat. Dogs are not given the leftover bones to gnaw on but are given off-cuts and offal. The dogs in Choijo's encampment scavenged on carcasses left by wolves but Saikhanaa would attempt to drive the dogs away from a carcass because he was told not to let the dogs drag the bones back to the encampment. Bones are left to disintegrate back into the pasture, eventually eroding back into the earth. Some bones, however, are given special significance.

The tibia bone of the sheep is a sacred object to Mongolian herders. When they are consuming a sheep, the choicest meat from the tibia bone is reserved for the eldest male of the family, who may then slice off small pieces of meat from the bone and pass them to other family members. A dried tibia bone may be placed in the central fire as a way to foretell future events. From the cracks that form on the bone, a herder who has learnt to read the appropriate signs may be able to predict where lost animals are located. According to Szynkiewicz (1990: 77), the tibia bone of the sheep signifies genealogical lineage or patrilineal descent for Mongolians. He gives an example of the significance of the tibia bone:

> On the third day after delivery the child is ceremonially washed in a broth made from the tibias and associated meat of a specially killed sheep. . . . Afterwards the [tibia] bone is tied to the cradle for a while and then preserved in the chest where the family valuables are kept and on which the family altar is placed. Beforehand, a *khadag* (ceremonial silk scarf), a sign of the highest honour, is attached to the bone. Each of the tibias stored in the chest should have a name, which is that associated with the child.

This description is interesting in that the tibia is designated as a sacred object by the attachment of a *khadag* and by giving the tibia itself a name that is associated with the child (see Chapter 4).

An exception to the utilisation of the animal for food is in the case of specific horses. Stallions and particularly fine racehorses are not eaten and the horse's head is placed next to a rock cairn (*övöö*) at the top of a nearby mountain peak and left to be picked clean by scavenging birds. In the case of Dogsomjav's encampment, Ochero places horse's skulls in the branches of a nearby lone tree so that wolves or dogs will not scavenge from the skulls and so that the bones will not be trampled into the ground. The essence of the horse is linked to the protective power, or "guardian spirit" of a high mountain peak, or large, lone tree. Throughout Altaic pastoral societies, in rites performed by shamans, male stallions and bulls, bears, deer, and eagles are regarded as having greater power than other beings (Tomikawa 2006). The stallion exhibits greater energy and power than other horses and this is reflected in the Mongolian herders' reverence of them (see Chapter 7).

THE *SETER* AS PROTECTOR

A blue sash (*khadag*) around a *seter* animal's neck is nicely representative of the intermixing of both Buddhism and shamanism. The blue sash signifies that a *seter*, or sacred animal, has been dedicated to *Tengger*. In Tibet, the most common silk sash used by Buddhist pilgrims as offerings is white (author's personal observation). In Mongolia it is always blue, representing the vastness of the Mongolian sky. The sky has great spiritual significance to Mongolians; according to shamanism, the heavens are the main spirit-being *Tengger* (or instead of shamanism, Tenggerism; see Bira 2004). According to Caroline Humphrey and Urgunge Onon (1996), the concept of *Tengger* is not precisely the same as the Western concept of sky, because *Tengger* means both the sky and the weather. The sky or weather acts with a will and has the power to retain balance in the world.

Onon emphasises that in shamanism it would be incorrect to say that everything has a soul but it does have its own "indeterminate energy." This energy is greater in some beings than others, particularly if they are unusual or lone objects that can be singled out, like a large, singular tree on a mountainside. In the case of the *seter*, its energy is far greater than other animals within the herd and can act as a protector against negative energy and harmful forces, such as extreme weather conditions or wolf attack. There is a saying that "if you hit one cow's horn, a thousand cow's horns will reverberate" (Humphrey with Onon 1996: 91). According to the same logic, if one individual is not killed, then this acts as protection for all the other members of the herd.

The following instance is a ceremony performed to recognise a *sarlag* cow as a *seter* animal. This is an account from my field notes, which involves a *seterlekh* ceremony conducted by a lama primarily to protect a woman who was seriously ill:

> The extended family invited a big, Mongolian-wrestling sized lama and his wife to stay overnight at the encampment, while on their journey to visit relatives. The only daughter of the patriarch of the encampment was ill with kidney problems, since the birth of her child the previous year. The lama arrived, opening up his leather satchel to reveal a couple of small bells and well-worn slips of paper with horizontal Tibetan text inscribed on them. He sat in impressive golden silk robes, near the calf pen, while the husband of the sick

> woman held a black-and-white yak still in front of the lama. The sick woman was absent but her father and her baby son witnessed the ceremony. The lama's chanting was incomprehensible to the herders (and no doubt to the yak), as it was in Tibetan, not Mongolian. But to all involved the intention was clear; this yak was singled out as a special individual and the ceremony was directed toward the cow. The lama asked the herder to say to the cow in Mongolian, "From this time onward, I will not kill you but we need milk from you and your new calves and fleece. You are the leader of our herd. You have to believe in us and always remain happy and healthy." A blue sash was tied around the cow's neck as a representation of her spiritual status. The yak was blessed with incense, then encouraged to drink milk from a bowl and drops were sprinkled along her head and back. When the lama had finished his incantations, he, his wife, and the family nonchalantly walked away, leaving the yak to shake the milk from her coat and give a few grunts. (Field note excerpt: anonymous summer encampment, August 24, 2005)

The lama had previously indicated to the family that the holy animal was to be a *sarlag* (or yak) cow that was young and dark in colour. Only Buddhist lamas that have studied specific Tibetan Buddhist texts are allowed to conduct a *seterlekh* ceremony. The herder is asked to speak directly to the cow, making a verbal pact that the family will not kill her and that she must aid as protector. Under everyday circumstances herders do not speak to individual herd animals in Mongolian but use calls and commands specific to the kind of animal (see Chapter 5). Because this individual has greater power than the average cow, it is assumed that she will understand her spoken instructions. Eduard Kohn relates similar circumstances with the Runa of Amazonia, in which a dog was given hallucinatory drugs so that the dog would be able to understand. In his words, "the Runa must make their dogs into shamans so that they can traverse the ontological boundaries that separate them from humans" (Kohn 2007: 13).

This individual cow functions as a protector of both the family and the rest of the cattle herd. When I asked a Buddhist lama from the nearby town, Chinzorig, about the *seter*'s ability to protect, he spoke of the *seter*'s powers of protection as all encompassing, because "the land, nature, and environment have the *seter* as protectors ... *seter* lead animals and all of nature, so actually the *seter* is responsible for protecting all kinds of life." When the big Buddhist monk who conducted the *seterlekh* ceremony visited the herding encampment, he also blessed a lone, sacred tree by reading sutra, and the family made offerings: They burned incense and butter lamps, filled bowls with candies and dairy products, and presented tiny bowls with wild berry juice, milk, distilled vodka, *airag*, and milk at the base of the tree. Once a year, herders come from great distances to make blessings to this big, lone tree in the valley. Similar to the *seter*, it is forbidden to touch the tree, or to chop the tree down for firewood, or timber. It is clear that the sacred tree and the *seter* cow have powers, beyond that of an ordinary cow or tree, which assist in protecting the herding family and their herd animals from harm. When a *seter* animal dies the herder takes the animal's body to the top of a high mountain, linking the essence of the *seter* animal with the powerful spirit of the mountain.

The horse holds great symbolic importance for Mongolians and is the most common animal designated as a *seter*, yet families can have a separate *seter* designated for each of their herds. In general the herders' attitude is to try to be egalitarian in the treatment of all the herd animals. When I would ask herding family members whether they have favourites, or special animals, they would say they did not have any. Dogsomjav responded to my question with a functional answer, saying "I like cows that produce a lot of milk and horses that are quiet but ultimately they will be ..." and she made a gesture indicating the motion for slitting open an animal's stomach when it is slaughtered. The *seter* is an exception in that it is singled out from the other members of the herd and is not eventually killed for food. Unlike a pet that experiences one-to-one interaction and special bonds with its human counterparts, herders usually *avoid* handling *seter* and essentially treat them as if they are beyond human intervention, similar to "wild" animals.

The circumstances surrounding the *seter* animal are not the same for every family. Choijo's family members milked their *seter* cow and clipped wool from it but would not kill the individual animal. When I asked why some herders handled the *seter* and some did not, herders invariably responded that "it depends on the family." Herders also informed me that it depends upon the specific instructions from individual monks.

The life of a herd animal can be interchanged with the death of a family member, providing a good example of the herders' reciprocal relationship with their herd animals. After a funeral, for the life that has been taken away from the herding family, another life may be "liberated," in the designation of a *seter* animal. In a similar reciprocal exchange, according to Rudenko (1970: 284), in the middle of the past century the death of a wealthy Kazakh resulted in a horse being granted "freedom" to counter the death of a herder:

> On the seventh day following the funeral people assembled and a great feast was offered to them. The saddle was put upside-down [or back-to-front?] on the dead man's horse and his clothes hung over it with hat on top, and the horse

Figure 10.2. A shaman's drawing of a *seter* animal. (Redrawn by the author from a pencil drawing in Dogsomjav's *ger*, Bulgan field site, 2005.)

was led in this manner to the dead man's tent, to which it was tethered by a lead rein. Then to the sounds of doleful singing the horse's tail was cropped. Such a horse was never ridden again and was called a widow.

The Animism of Objects

Dogsomjav's extended family does not have living *seter* animals. Instead they have a pictorial representation of a *seter* to act as protection for the herd. The *seter* picture is displayed behind glass, next to the small Buddhist shrine toward the rear of the *ger* (see Figure 10.2). Dogsomjav's husband had formally been given a *seter* drawing from his mother, for whom it had been created by a knowledgeable shaman woman (or *udgan*). However, when Dogsomjav's husband died, a lama told the family that this *seter* picture was to be destroyed and thrown into the river, as the picture had brought bad luck upon the family. Another picture was drawn and kept by Dogsomjav in her *ger*. As seen in Figure 10.2, the picture depicts other items that are important in a *seterlekh* ceremony in the form of milk, smelling salts, incense (an aromatic pine called *arts*), and candles burning animal fat. It is significant that a symbolic drawing of the holy animal is enough to protect the rest of the herd, provided that a lama has blessed the picture. When I asked why the family did not select a real individual from the herd, Dogsomjav told me that it is too risky. If anything negative happens, for instance, if the animal is attacked and eaten by wolves, it brings bad luck upon the entire herd. A pictorial representation avoids the risk of the occurrence of bad omens and means that all animals can be utilised for food or other purposes. The power of the *seter* animal is still evident even when represented as an inanimate drawing.

Dogsomjav also had different means of protecting the herd from harm, not relying on the power of the symbolic sacred animal alone. A powerful object made from the neck vertebrae of a wolf and tied with coloured cloth hangs on the wall beside a wooden stick used for sprinkling the first milk tea of the day (*tsatsal*). The object is used as protection to ward off wolves from attacking the herd animals. A herding family may ask visiting lamas or old herders who can decipher Tibetan to read specific sutra while burning incense and to walk amongst the herds chanting in order to protect the herds from harm, from becoming sick or injured, from being stolen, or from being attacked by wolves.

Herders believe that some objects have power and have an ability to influence events in a positive or negative way, but this is not the case with all inanimate things (also see Humphrey with Onon, 1996). Natural features such as a particularly notable mountain, cliff, large outcrop, or lone tree are recognised as sacred. Those objects that have greater significance and symbolic meaning have a blue silk scarf (*khadag*) tied to one end; just as a *khadag* is tied around the sacred animal's (*seter*'s) neck, a *khadag* is also placed on an *övöö* on a mountaintop, or tied around the trunk of a lone tree. Specific objects such as a knife, horse scraper (*khusuur*), branding iron (*tamga*), or horse-head fiddle (*morin khuur*) are examples of objects with power. Disposable objects like a hide, food, leather ties, or the felt inside a boot would not require the same degree of respect. Objects that require skill on the part of the user and that are an integral part of a herder's existence are highly valued, such as the lasso-pole, or *uurga*, which is necessary for catching a horse. A herder is expected to treat the *uurga* with respect, carrying it in a specific way under one arm; when not in use the *uurga* should not be left lying on the ground where it could be stood upon. If someone inadvertently stands upon an *uurga*, this is thought to result in bad consequences for that person. In accordance with animist systems of belief, Mongolian herders attribute power not only to animate beings but to what we Westerners perceive as inanimate objects and natural features.

CONCLUSION

Different practices and beliefs provide a powerful window into a people's attitude toward their herd animals and their ritualised animal husbandry practices. This is related to conceptions of death, particularly perspectives on killing animals. It is evident from Mongolian herders' conduct surrounding the slaughter of their animals that they demonstrate a reciprocal relationship with their herd animals. This derives from an animist system of belief in reincarnation and continuity rather than a Western view of domination based on monotheistic principles described in the Old Testament.

Mongolian herders believe that if a herder behaves in an inappropriate manner, his or her actions will have a direct effect on the health and prosperity of the herd animals and therefore on the herding family as a whole. Each herder is expected to treat the animals as persons who require care and respect, particularly because a herder could become one of the five kinds of animal (*tavan khoshuu mal*) in the next life. One means by which the moral dilemma of having to kill other lives is solved is by designating an animal as sacred, as a *seter* animal, which will not be killed. The sacred animal then acts as protector for the rest of the herd, the family, and even the surrounding landscape. Just as the *seter* is singled out from the rest of the herd as spiritual protector, so too are other features of the landscape, such as an unusually shaped mountain, or the only large, lone tree within a river valley. This mountain, or tree, is blessed by local herders and it too acts as a protector. The lone tree (*lus savdag*) exemplifies plants, and the mountain peak with a rock cairn (*övöö*) exemplifies geophysical entities (Humphrey with Onon 1996). Similarly, the *seter* exemplifies the animals within a herd, with the power to protect them from harm. A Mongolian herder is not a being who holds great power over other beings; instead herder and herd animal have a mutual co-dependence upon one another that entails trust, respect, and reciprocity.

WINTER

Plate 25. Sacred lone tree in the valley.

Plate 26. Boy amongst sheep and goats.

Plate 27. Early lamb with children.

Plate 28. Snow falling inside a *ger*.

Plate 29. Passing snuff, with the horse herd in the background.

Plate 30. Moving a sick sheep under shelter.

Plate 31. Herding sheep and goats in winter.

Plate 32. Footprints leaving a neighbouring encampment.

CONCLUSION

CO-DOMESTIC LIVES

What is animal "domestication" in Mongolia? On the basis of my experience living in the Khangai Mountains, I came to the view that Mongolian herding families and herd animals have developed co-domestic existences with one another. In adaptive terms, I suspect that both human and non-human animals have experienced symbiosis, or a co-evolutionary domestication process, through their profound influence upon one another over thousands of years. This long-term perspective, however, was not the primary focus of this book. Instead I have described a window into the herder–herd animal relationship in two valleys in the Khangai Mountains, within the lifetimes of specific individuals. My intention was to provide examples of how human and non-human live together through intercultural, interspecies existences in Mongolia.

Mongolian herders and herd animals rely on one another in a reciprocal relationship of co-dependence. While I was in the field, I realised how important the herd animals were to the herding families' everyday lives and ultimately for their survival. Herders refer to the co-domestic herd animals simply as the five kinds of animal, or *tavan khoshuu mal* (horses, cattle, sheep, goats, and camels), who are nurtured by and regularly engage with humans within the co-domestic sphere of the herding encampment. As Naraa succinctly put it, "We feed them and they feed us."

This mutual relationship is achieved through the following means: Herders allow the herd animals to retain their inherent social structure, which is similar to the herding families' own social structure; the two groups engage with one another in a cross-species ecosocial landscape, specifically the co-domestic sphere of the herding encampment (*khot ail*); herders name individual animals and recognise these animals as persons with individual characteristics and personalities; herder and herd animal enculturate each other through the use of species-specific vocalisations and body language as a means of communication; and herders habituate the herd animals to humans from a young age, encouraging a lack of fear and a relationship of trust. Just as a Mongolian stallion is accepted as leader by mares and young, herders are accepted by the herd animals as members of the herd, whereby the herder similarly takes the role of leader. Humans and non-humans co-exist within a cross-species social structure and hierarchy, forming separate but interacting herd societies.

The herder–herd animal co-domestic relationship has a profound influence on Mongolian herding life. This can be illustrated by the multilayered connection between human and horse, which has produced a reverence for the horse as a key symbol and a constitutive participant in Mongolian society. Horses are therefore pivotal in all major contexts of the herding existence: in social relationships; subsistence and mobility; artistic performance; competition in the form of horseracing; and in death, spirituality, and the afterlife. A herder's whole cosmology and world view encompass other beings. Humans and non-humans are engaged in a constant cycle of death and rebirth through reincarnation. The expectation is that positive and respectful behaviour towards other beings will be reciprocated in kind. A fascinating example of this is the designation of a *seter*, or sacred animal within a herd. The *seter* as an individual exemplifies the animals within a herd, containing heightened powers to protect the herding family and other herd animals from harm. In exchange, the *seter* is never killed or eaten by the herding family. The Mongolian herder's world view is essentially animist, similar to Northern Arctic hunting peoples, as herders recognise the herd animals they associate with daily as persons to be regarded with respect in life and in death.

Life as a herder is structured around and mapped onto the lives of the herd animals. The herding family wakes when the cows are ready to be milked and retires to bed only when the sheep and goats have been herded safely into the pens and corrals for the night. A herder's daily tasks coincide with the gender and age of the animals: Young children are encouraged to herd the young animals near the encampment; strong, young men handle the heavier male geldings and oxen; women nurture young and milk the female animals. When the seasons change and the herd animals begin to search for new pastures, herding families take their cue and move on to the next seasonal encampment.

Mongolian herders are a part of the landscape. There is not a distinctive dividing line between where nature begins and culture ends. To be a co-domestic animal is to engage with others, human or non-human, and to be encompassed within the landscape, or co-domestic sphere of the encampment. The co-domestic sphere is "home" for herders and herd animals alike. Beyond this sphere, other beings are not as interlinked with humans but still have the potential to be strong and powerful influences. Wild medicinal plants have the power to counteract illnesses suffered by the herding family and herd animals. The wolf is regarded with ambiguity, as the wolf is capable of invading the co-domestic sphere. A wolf will be shot on sight to defend the herd animals against attack, yet the wolf is also viewed by the herder as containing strength and power. Clearly, it is not just herders who constitute the Mongolian herding existence. Herd animals, other beings, and the landscape are all actors in, and an active part of, the sociality of herding life.

The traditional Western perspective on domestication presupposes that a herd animal succumbs to domestication at the whims of a superior human through intentional design, in a relatively mechanistic relationship designed to maximise human profit for human gain. By contrast, Mongolian herder and herd animal live co-domestic existences together, their lives inextricably interlinked and dependent upon one another. Herd animals, and herders, are usually born within the encampment (*khot ail*); learn each other's habits, temperaments, and behaviours from an early age; interact with one another on a daily basis; and then die within this same sphere as co-adapted co-inhabitants, who are co-domestic with one another.

EPILOGUE

Life in Mongolia is not easy. The temperatures are highly changeable and the weather conditions are unrelentingly extreme. In its stark and incredible beauty, the Mongolian landscape left a deep impression upon me. I felt that the herding families I lived with, the herd animals I came to know, and the now-familiar landscape became a part of me as a person. Mongolian herders are hardy; they approach life with strength and independence but with an unassuming humour and humility. The people I met were wonderfully accepting and generous towards me as "other" and accepted me as part of their extended family. This is an attitude that is reflected in their acceptance of others in general, both human and non-human. I knew on the day that I left Dogsomjav's encampment, while peering through the dusty window of the rickety jeep at the now-familiar valley, that Mongolia was now a part of me and I would be compelled to return.

I did indeed return, this time for a short follow-up trip in the spring of 2007. I found that life had changed for the members of both Choijo and Dogsomjav's encampments in the fifteen months since I had left.[1] When

[1] I relate here the changes that occurred within Dogsomjav's encampment, although circumstances had also changed considerably within Choijo's encampment and for members of neighbouring herding encampments.

I arrived at the Bulgan field site, Dogsomjav's *ger* was still at her well-established winter site and she was busily tending the cows that had given birth to calves that season. Much of her time was also occupied in looking after her grandson Enkhee, who was born while I was living in the encampment but had since developed into a troublesome two-year-old for the now tired-looking Dogsomjav.

Members of Ochero's family had moved to their spring camp, approximately a kilometre away, to provide a more suitable location for the sheep and goats with their young. They had temporarily combined the sheep and goat herd with Sanjaa's family's herd (Sanjaa is Boloro's brother), as he had split away from his former encampment, taking his family's herd animals with him. Their contribution and cooperation was welcome because the more hands available to help out, the easier herding (and milking) becomes. Ochero's son, Öskö, was now six and regularly herding on horseback. I could hear him practicing his herding vocalisations, even holding the elongated pole (*shelbuur*) in one hand, as he rode around on the *nomkhon* gelding, Bambaruushka. Öskö was now adept at recognising all the sheep and goats, including their young, and regularly checked their dust-filled eyes to see whether they had become infected. Each morning I saw him riding off down the valley for lessons from a local kindergarten teacher seven kilometres away, without anyone to accompany him except the horse he was riding.

Lhagva, her husband, and two children had moved to the provincial centre of Erdenet. They had moved so that their daughter, Undera, could attend school. This was a big setback for Dogsomjav, as she had relied on Lhagva's industrious help and support. It was disheartening to find someone as knowledgeable in treating illnesses using medicinal herbs and as dedicated to her cows as Lhagva having to work long hours in a Japanese-owned pet food factory. Lhagva's family could no longer rely on the meat and dairy products from their herd animals alone. A monetary income is necessary in the Mongolian urban environment to pay for food, rates, water, and electricity, none of which are needed if one is living in the countryside. When I asked Lhagva previously in 2005 about life in the countryside, she emphasised to me how she would rather be where she was than in the city. Previously she had said, "I like to live in the countryside because I'm so busy here. In the city it's very boring for me, just watching television and not doing much else. Work here in the countryside is good; I like it." In the countryside Dogsomjav's encampment is relatively self-sufficient, relying mainly on gifts and exchange with their extended family in the city for outside necessities. Dogsomjav's encampment may have no guaranteed monetary income but the quality of life is good, the work is varied, and there is a strong sense of community and sociality amongst herders and herd animals.

Circumstances had also changed for the herd animals. Dogsomjav and Lhagva enthusiastically updated me on the changes that had occurred with the various animals, telling me about their lives just as they told me about the herders I knew. The Black Stallion (*Khar Azarga*) had been given to another herder and although he had already been in the role of stallion at Dogsomjav's encampment, he was now the stallion of another band (see Chapter 4). A young white horse (*Bor Azarga*) was the stallion that replaced him, but this stallion was killed by wolves while he was attempting to protect the mares and young from attack. The Big Red (*Tom Ulaanaa*) ox damaged his leg and began to lose weight, so he was exchanged with another herder for some younger cattle. Three of the cattle that I had seen born as calves in 2005 were now castrated and would be trained as oxen to pull an ox cart the following year. The cow Big White Eyebrows (*Tom Khomsog tsagaan*) had complications giving birth the previous year, so she was eaten in the autumn cull. Her daughter, who was the firstborn calf in 2005 (*Uuganaa*), was now a young mother herself and was being trained by Dogsomjav to the milking routine (see Chapter 6). The calf Birthmark (*Menget*) was given away by Dogsomjav as a gift to another herder.

Clearly, time was not static for Dogsomjav's encampment. Extended family dynamics and the composition of the herds had changed (particularly with the injection of key breeding males from other herds). At the same time I was struck by the fact that life within the encampment remained much the same, and that the daily and seasonal cycle of events were progressing as usual. The level of social engagement between herders and herd animals had also not changed. Most of the herders' focus during my visit was in matching mother sheep and goats with their young for nursing each morning and evening; or with the evening's cow milking and getting the new calves habituated to the milking routine. I slipped back into the slow pace of life within the encampment and into the modes of engagement with the herding family and the herd animals as if I had never left. Life was cycling around again, just as herders habitually circumambulate around the inside of their *ger*, a sacred cairn (*övöö*), or around the hitching posts (*mornii uyaa*). The regeneration of life was taking place before my eyes as the last of the year's calves were born and the first rains of spring encouraged new shoots of green vegetation to sprout up on the valley floor. Individual herders and herd animals were born, and some had died (or left the encampment), but life continued on, reincarnated and revitalised with the warming of spring. Together herders and herd animals were facing and engaged with the cycle of life in the Khangai Mountains of Mongolia.

APPENDIX

Table A.1. Categorisation of horse colours in Mongolia

Colour	Description	Translation
WHITE (*Tsagaan*)	**White (eyes are black)**	***Tsagaan***
	White with brownish tinge (but the word means "brown")	*Bor*
	Mottled white	*Tsookhor Tsagaan*
	White with mixture of colour	*Zagal*
	Dark–white mixture	*Khüren zagal*
	Brownish-white mixture	*Kheer zagal*
	Roan (speckled grey)	***Buural***
	Blue–grey roan	*Khökh buural*
	Strawberry roan	*Ulaan buural*
	Brownish-grey	*Khüren buural*
	Pinkish-grey	*Yagaan buural*
	Grey (ashen)	***Saaral***
	Dapple grey	*Khökh saaral*
	Iron grey	*Khökh bor*
	Greylag goose–crane grey	*Kheeriin galuu*
	Whitish-blue	*Tsenkher*
	Dark grey	*Khar saaral*
	Flea-bitten–spotted grey	*Bidertei saaral*
	Grey-ticked–mottled pink	*Yagaan tsookhor*
	Dun (Isabella)	***Khaliun***
	Light–yellowish dun	*Bor khaliun*
	Blue–greyish dun	*Tsaivar khaliun*
	Whitish-dun	*Tsagaan khaliun*
	Creamy dun	*Shar khaliun*
	Dark brown dun	*Kheer khaliun*
	Light-bay dun	*Khongor khaliun*
	Cream	***Sharga***
	Reddish-cream	*Ulaan shar*
	Sorrel	*Shar kheer*
	Palomino (golden, white mane and tail)	*Tsavdar*

(continued)

Table A.1 *(continued)*

Colour	Description	Translation
BLACK (*Khar*)	**Black**	***Khar***
	Velvet black	*Khilen khar*
	Dull black	*Khar saaral*
	Glossy black	*Gyaltalzcan khar*
	Glistening–shiny black	*Gölchgör khar*
	Light black (or dark chestnut, also called "green")	*Khar kheer*
	Spotted black	*Khar khaltar*
	Black piebald	*Khar alag/khalag*
	Bay roan	*Buural kheer*
	Dark chestnut bay–dirty bay	*Kheer khaltar*
	Bay with white patches	*Kheer alag*
CHESTNUT (*Zeerd*)	**Chestnut (golden red)**	***Zeerd***
	Reddish chestnut	*Ulaan zeerd*
	Reddish-yellow chestnut	*Tsavidar zeerd*
	Dark chestnut	*Khuren zeerd*
	Chestnut roan	*Ulaan buural*
	Dark brown–weasel	***Khüren***
	Brown piebald	*Khüren alag*
	Ebony brown	*Khüren khar*
	Bluish brown	*Tsenkher khüren*
	Brown bay	*Bor kheer*
	Dirty	*Khaltar*
	Bay (brown, black mane and tail)	***Kheer***
	Red bay	*Ulaan kheer*
	Tan–fawn bay	*Khul kheer*
	Bay roan	*Buural kheer*
	Dark chestnut–dirty bay	*Kheer khaltar*
	Piebald (bay with white patches)	*Kheer alag*
	Skewbald (tan with white patches)	*Khul alag*
LIGHT BAY (*Khongor*)	**Light bay**	***Khongor***
	Entirely light bay	*Ukhaa khongor*
	Wild-type	*Khulan*
	Fawn	*Khul*
	Skewbald	*Khul alag*
BODY MARKINGS	**Mottled (spots or marks)**	***Tsookhor***
	Mottled brown	*Khüren tsookhor*
	Mottled red	*Yagaan tsookhor*
	Mottled piebald	*Alag tsookhor*
	Piebald (patches)	*Alag*
	Dappled	*Khölögchin*
	Spotted	*Tolbotoi/bidertei*
	Mottled white markings	***Tsookhor tsagaan***
	White below fetlocks	*Sagag tsagaan*
	White muzzle	*Khamar tsagaan*
	White blaze (or "bald")	*Khalzan*
	White sock	*Bilevsen*
	White moon	*Sartai*
	White eye	*Tsekher*

Table A.2. Medicinal plants used to treat illnesses, particularly in relation to herd animals

Name	Description of part used	Treatment
Agi	Grey-green coloured and scented	Provides vitamins for tired animals
Altan gagnuur (*Rhodiola quadrifida*, mountain welder):	Use of greenish-yellow flower	The rhizome is used to treat animals (incl. humans) with broken limbs
Borolzgono	Slender, brown tree with yellow flowers and very little, red berries	Good for people and animals with diarrhoea or when tired; sometimes drunk with tea
Dagsh	Green pine, looks like leaves; oily	Contains vitamins for tired animals
Ganga (wild thyme)	Perfumed plant	Good for animals with a high temperature, for colds, or for tired animals, esp. cows after giving birth.
Khalgai (*Olslig khalgai*, *Urtica cannabina*, nettle)	A type of tall, green thistle; many plants at Dogsomjav's summer campsite	Many vitamins; good for tired animals, colds, and keeping blood pressure low; collected and dried to feed to cows as supplementary feed
Khar modnii möög buyou agal (black wood/tree mushroom)	Collected from trees in the forest on horseback; the higher ones are better than those close to the ground	Good for wolf attack, after cows have given birth, and eaten as food
Mangir (or *Mongol congino*, *Allium mongolicum*, Mongolian onion)	Flower is white-pink	Given to animals when tired in spring
Söd	Use the pods on the red seed head	Good for stomach ache and for calves, lambs, and kids with diarrhoea from wet ground in spring
Songino (*Altain songino*, *Allium altaicum*, onion)	Small onion, produces white flower	Used for tired animals
Tsagaan tseene (*Paeonia lactiflora*)	White flower	The roots are used to treat animals when tired in spring
Yargui	Strong smelling, purple flower; one of first to appear in spring	Consumed by herd animals while grazing in spring; provides energy and vitamins; effective against parasites
Yaz khüü	Flower found in grassland	Dried, then boiled, and placed on wound after horses and cattle have been attacked by wolves

(continued)

Table A.2 *(continued)*

Name	Description of part used	Treatment
Yum düüjin (*Dianthus superbus*)	Long, slender stem with purple flower.	Plant is boiled; good for the uterus; given to cows after birth to release the placenta
Zeergene (*Khonin zeergene, Ephedra sinica, Chines Ephedra*)	Produces small, red berries and yellow flowers; grows as a low plant; looks like bracken	The shoots have a bitter taste; good for kidney diseases as a wound-healing remedy; sheep and goats like eating this plant

Note: These are the plants collected and used by Dogsomjav's encampment in 2005. Where plant names do not have the Latin name, this is because of the rarity of the plant or because the herders have a common name that does not align with the recognised Mongolian nomenclature for the plant.
Source: Mongolian proper names and Latin names for plants were also attained from Galbadrakh and Ganchimeg (2000) and Ligaa and Tsembal (2003).

REFERENCES

Aikio, P.

1989 "The Changing Role of Reindeer in the Life of the Sami." In *The Walking Larder: Patterns of Domestication, Pastoralism and Predation*, ed. J. Clutton-Brock (pp. 169–183). London: Unwin Hyman.

Altangerel, D.

1987 *How Did the Great Bear Originate? Folk-Tales from Mongolia*. Ulaanbaatar: State Publishing House.

2006 *Mongolian Folktales and Legends*. Ulaanbaatar: SOD Press.

Amoroso, E. C., and P. A. Jewell.

1963 "The Exploitation of the Milk-Ejection Reflex by Primitive Peoples." In *Man and Cattle: Proceedings of a Symposium on Domestication*, ed. A. E. Mourant and F. E. Zeuner (pp. 126–137). London: Royal Anthropological Institute of Great Britain and Ireland.

Anderson, D. G.

2000 *Identity and Ecology in Arctic Siberia: The Number One Reindeer Brigade*. Oxford: Oxford University Press.

Anderson, M.

1986 From Predator to Pet: Social Relationships of the Saami Reindeer-Herding Dog. *Central Issues in Anthropology* 6(2): 3–11.

Andrews, R. C.

1932 *The New Conquest of Central Asia: A Narrative of Explorations of the Central Asiatic Expeditions in Mongolia and China, 1921–1930* (Vol. 1). New York: American Museum of Natural History.

Anthony, D. W.

2007 *The Horse, the Wheel and Language: How Bronze-Age Riders from the Eurasian Steppes Shaped the Modern World*. Princeton, NJ: Princeton University Press.

Atwood, C. P.

1996 Buddhism and Popular Ritual in Mongolian Religion: A Reexamination of the Fire Cult. *History of Religions* 36(2): 112–139.

Azzaroli, A.
1985 *An Early History of Horsemanship*. Leiden: E. J. Brill.

Baabar.
2004 *History of Mongolia*. Cambridge: Whitehorse Press.

Balzar, M. M.
1990 *Shamanism: Soviet Studies of Traditional Religion in Siberia and Central Asia*. New York: M. E. Sharpe.
1996 Flights of the Sacred: Symbolism and Theory in Siberian Shamanism. *American Anthropologist* 98(2): 305–318.

Banks, M.
2001 *Visual Methods in Social Research*. London: Sage.

Barfield, T. J.
1993 *The Nomadic Alternative*. Englewood Cliffs, NJ: Prentice-Hall.

Barth, F.
1961 *Nomads of South Persia: The Basseri Tribe of the Khamseh Confederacy*. Oslo: Universitete Etnograpfisko Museum, University of Oslo.

Bateson, G.
1972 *Steps to an Ecology of Mind*. New York: Ballantine Books.

Batsukh, G., and E. Zagdsuren.
1991 *Animal Genetic Resources*. Ulaanbaatar: Food and Agriculture Organisation.

Bawden, C. R.
1968. "Mongol Notes: Some 'Shamanist' Hunting Rituals from Mongolia." *Central Asiatic Journal* 12(2): 101–143.

Bayarsaikhan, B.
2006 *Travelling by Mongolian Horse*. Ulaanbaatar: Interpress Co.

Becker, J.
1992 *The Lost Country: Mongolia Revealed*. London: Sceptre.

Belyaev, D. K.
1979 "Destabilizing Selection as a Factor in Domestication." *Journal of Hereditary* 70: 301–308.

Belyaev, D. K., I. Z. Plyusnina, and L. N. Trut.
1984–1985 "Domestication in the Silver Fox (*Vulpes fulvus* Desm): Changes in Physiological Boundaries of the Sensitive Period of Primary Socialization." *Applied Animal Behaviour Science* 13: 359–370.

Benedict, B.
1969 "Role Analysis in Animals and Men." *Man* 4(2): 203–214.

Bira, S.
2004 "Mongolian Tenggerism and Modern Globalism: A Retrospective Look on Globalisation." *Journal of the Royal Asiatic Society* 3(14): 3–12.

Bird-David, N.
1990 "The Giving Environment: Another Perspective on the Economic System of Gatherer-Hunters." *Current Anthropology* 31(2): 189–196.

1999 "'Animism' Revisited: Personhood, Environment, and Relational Epistemology." *Current Anthropology* 40: S67–S91.

Bjørnstad, G., N. Ø. Nilsen, and K. H. Røed.

2003 "Genetic Relationship Between Mongolian and Norwegian Horses?" *Animal Genetics* 34: 55–58.

Bökönyi, S.

1984 "Horse." In *Evolution of Domesticated Animals*, ed. I. L. Mason (pp. 162–173). London: Longman.

1989 "Definitions of Animal Domestication." In *The Walking Larder: Patterns of Domestication, Pastoralism and Predation*, ed. J. Clutton-Brock (pp. 22–27). London: Unwin Hyman.

Bold, A.

2003 "Mongolian Birds." In *Mongolia Today: Science, Culture, Environment and Development*, ed. D. Badarch, R. A. Zilinskas, and P. J. Balint (pp. 143–171). London: RoutledgeCurzon.

Bold, B-O.

2001 *Mongolian Nomadic Society: A Reconstruction of the 'Medieval' History of Mongolia*. Richmond, Surrey: Curzon Press.

Boyd, L., and K. A. Houpt.

1994 *Przewalski's Horse: The History and Biology of an Endangered Species*. New York: State University of New York Press.

Bruford, M. W., D. G. Bradley, and G. Luikart.

2003 "DNA Markers Reveal the Complexity of Livestock Domestication." *Nature* 4: 901–910.

Bulliet, R. W.

2005 *Hunters, Herders and Hamburgers: The Past and Future of Human–Animal Relationships*. New York: Columbia University Press.

Buxton, L. H. D.

1926 "The Inhabitants of Inner Mongolia." *The Journal of the Royal Anthropological Institute of Great Britain and Ireland* 56: 143–161.

Byambasuren, D., and F. Luigi.

2004 *The Story of the Weeping Camel* (film). Germany–Mongolia: G. Schech Filmproduktion and ThinkFilm.

Bynie, B.

2002 *The Country Report on Animal Genetic Resources*. Ulaanbaatar: National Consultative Committee on Animal Genetic Resources of Mongolia.

Campbell, A.

1995 "Animism Revisited: Personhood, Environment, and Relational Epistemology." In *Readings in Indigenous Religions*, ed. G. Harvey (pp. 73–105). London and New York: Continuum.

Campbell, B.

2005 "On 'Loving Your Water Buffalo More than Your Own Mother': Relationships of Animal and Human Care in Nepal." In *Animals in Person*, ed. J. Knight (pp. 79–100). Oxford: Berg.

Carruthers, D.

1913 *Unknown Mongolia: A Record of Travel and Exploration in North-West Mongolia and Dzungaria*. Vol. 1. London: Hutchinson & Co., London.

Cassidy, R.

2002 *The Sport of Kings: Kinship, Class and Thoroughbred Breeding in Newmarket*. Cambridge: Cambridge University Press.

2007 "Introduction: Domestication Reconsidered." In *Where the Wild Things are Now: Domestication Reconsidered*, ed. R. Cassidy and M. H. Mullin (pp. 1–26). Oxford: Berg.

Clutton-Brock, J.

1987 *The Natural History of Domesticated Mammals*. Cambridge: Cambridge University Press.

1989 *The Walking Larder: Patterns of Domestication, Pastoralism and Predation*. One World Archaeology Series. London: Unwin Hyman.

1992 *Horse Power: A History of the Horse and the Donkey in Human Societies*. London: Natural History Museum Publications.

1994 "The Unnatural World: Behavioural Aspects of Humans and Animals." In *Animals and Human Society: Changing Perspectives*, ed. A. Manning and J. Serpell (pp. 23–35). New York: Routledge.

Clutton-Brock, T. H., and J. M. Pemberton.

2004 *Soay Sheep: Dynamics and Selection in an Island Population*. Cambridge: Cambridge University Press.

Cooper, J. J., C. Ashton, S. Bishop, R. West, D. S. Mills, and R. J. Young.

2002. "Clever Hounds: Social Cognition in the Domestic Dog (*Canis familiaris*)." *Applied Animal Behaviour Science* 81: 229–244.

Coppinger, R., and R. Schneider.

1995 "Evolution of Working Dogs." In *The Domestic Dog: Its Evolution, Behaviour and Interactions with People*, ed. J. Serpell (pp. 22–47). Cambridge: Cambridge University Press.

Damdinjaviin, D.

2005 *Tumen Shinjit Gaikhamshigt Khurdan Morid* [Ideal and Marvellous Racehorses]. Ulaanbaatar: No publisher noted.

Darwin, C.

1868 *The Variation of Animals and Plants under Domestication*, Vols. 1 and 2. London: John Murray.

Dawkins, M. S.

2006 "Through Animal Eyes: What Behaviour Tells Us." *Applied Animal Behaviour Science*. 100(1–2): 4–10.

Dawkins, R.

1978 "Animal Signals: Information or Manipulation?" In *Behavioural Ecology: An Evolutionary Approach*, ed. J. R. Krebs and N. B. Davies (pp. 380–402). Oxford: Blackwell.

1982 *The Extended Phenotype: The Gene as the Unit of Selection*. Oxford: Freeman.

Dawkins, R. and J. R. Krebs.
1979 "Arms Races Between and Within Species." *Proceedings of the Royal Society of London, Series B.* 205: 489–511.

Dennet, D. C.
1995 *Darwin's Dangerous Idea.* New York: Simon & Schuster.
1996 *Kinds of Minds: Towards an Understanding of Consciousness.* New York: HarperCollins.

Descola, P.
1992 "Societies of Nature and the Nature of Society." In *Conceptualizing Society*, ed. A. Kuper (pp. 107–126). London: Routledge.
1994 *In the Society of Nature: A Native Ecology in Amazonia.* New York: Cambridge University Press.
1996 "Constructing Natures: Symbolic Ecology and Social Practice." In *Nature and Society: Anthropological Perspectives*, ed. P. Descola and G. Palsson (pp. 82–102). London: Routledge.

Despret, V.
2004 "The Body We Care For: Figures of Anthropo-Zoo-Genesis." *Body and Society* 10(2–3): 111–134.

Diamond, J. M.
1966 "Zoological Classification System of a Primitive People." *Science* 151: 1102–1104.
1997 *Guns, Germs, and Steel: The Fates of Human Societies.* New York: Norton.

Diamond, J. M., and P. Bellwood.
2003 "Farmers and Their Languages: The First Expansions." *Science* 300(25): 597–603.

Dierendonck, M. C. van, and D. Goodwin.
2005 "Social Contact in Horses: Implications for Human-Horse Interactions." In *The Human-Animal Relationship*, ed. F. N. de Jong and R. van den Bos (pp. 65–81). Animals in Philosophy and Science, Vol. 4. Assen, Netherlands: Van Gorcum.

Dorjgotov, H.
1999 *English–Mongol, Mongol–English Animal Husbandry Glossary.* Ulaanbaatar: T & U Printing.

Downs, J. F., and R. B. Ekvall.
1965 "Animals and Social Types in the Exploitation of the Tibetan Plateau." In *Man, Culture and Animals: The Role of Animals in Human Ecological Adjustments*, ed. A. Leeds and A. P. Vayda (pp. 169–184). Washington, DC: American Association for the Advancement of Science.

Ducos, P.
1989 "Defining Domestication: A Clarification." In *The Walking Larder: Patterns of Domestication, Pastoralism and Predation*, ed. J. Clutton-Brock, trans. M. Matthews (pp. 28–30). London: Unwin Hyman.

Dunlop, S.
2000 *Conquest and Change: Mongol Herding in Xilingol, Inner Mongolia.* PhD dissertation, Australian National University, Canberra.

Dwyer, P. D.
1996 "The Invention of Nature." In *Redefining Nature: Ecology, Culture and Domestication,* ed. R. Ellen and K. Fukui (pp. 157–186). Oxford: Berg.

Dwyer, P. D., and M. Minnegal.
2005 "Person, Place or Pig: Animal Attachments and Human Transactions in New Guinea." In *Animals in Person,* ed. J. Knight (pp. 37–60). Oxford: Berg.

Dyson-Hudson, R., and N. Dyson-Hudson.
1980 "Nomadic Pastoralism." *Annual Review of Anthropology* 9: 15–61.

Ekvall, R. B.
1961 "The Nomadic Pattern of Living among the Tibetans as Preparation for War." *American Anthropologist* 63(6): 1250–1263.
1968 *Fields on the Hoof: Nexus of Tibetan Nomadic Pastoralism.* New York: Holt, Rinehart & Winston.

Ellen, R., and K. Fukui.
1996 *Redefining Nature: Ecology, Culture and Domestication.* Oxford: Berg.

Epstein, H.
1969 *Domestic Animals of China.* Farnham Royal: Commonwealth Agricultural Bureaux.

Erdenebaatar, B.
1996 "Socio-Economic Aspects of the Pastoral Movement Patterns of Mongolian Herders." In *Culture and Environment in Inner Asia, Vol. 1: The Pastoral Economy and the Environment,* ed. C. Humphrey and D. Sneath (pp. 58–108). Cambridge: The Whitehorse Press.

Erdenekhuyag, D.
2004 *The Horses of Chinggis Khan.* Ulaanbaatar: Hiimori Printing.

Eriksen, T. H.
2001 *Small Places, Big Issues: An Introduction to Social and Cultural Anthropology.* London: Pluto Press.

Evans-Pritchard, E. E.
1939. "Nuer Time-Reckoning." *Africa* 12(2): 189–215.
1940 *The Nuer: A Description of the Modes of Livelihood and Political Institutions of a Nilotic People.* Oxford: Oxford University Press.
1956 *Nuer Religion.* Oxford: The Clarendon Press.

Even, M. -D.
1991 "Shamanism of the Mongols." In *Mongolia Today,* ed. S. Akiner (pp. 183–205). London: Kegan Paul.

Exupéry, A. de Saint.
1943 *The Little Prince.* Translated from the French by K. Woods. Paris: Gallimard.

Feeny, D., F. Berkes, B. J. McCay, and J. M. Acheson.
1990 "The Tragedy of the Commons: Twenty-Two Years Later." *Human Ecology* 18: 1–19.

Fernández-Giménez, M. E.
1999 "Reconsidering the Role of Absentee Herd Owners: A View from Mongolia." *Human Ecology* 27(1): 1–27.
2002 "Spatial and Social Boundaries and the Paradox of Pastoral Land Tenure: A Case Study from Postsocialist Mongolia." *Human Ecology* 30(1): 49–77.

Fijn, N.
2007 "Filming the Significant Other: Human and Non-Human." *The Asia Pacific Journal of Anthropology* 8(4): 297–307.

Finch, V. A., and D. Western.
1977 "Cattle Colors in Pastoral Herds: Natural Selection or Social Preference?" *Ecology* 58(6): 1384–1392.

Fratkin, E.
1997 "Pastoralism: Governance and Development Issues." *Annual Review of Anthropology* 26: 235–261.

Fuentes, A.
2006 "The Humanity of Animals and the Animality of Humans: A View from Biological Anthropology Inspired by J. M. Coetzee's Elizabeth Costello." *American Anthropology* 108(1): 124–132.

Fukui, K.
1979 "Cattle Colour Symbolism and Inter-Tribal Homicide among the Bodi." In *Warfare among East African Herders*, ed. K. Fukui and D. Turton (pp. 147–177). Senri Ethnological Studies, No. 3. Osaka: National Museum of Ethnology.
1996 "Co-evolution Between Humans and Domesticates: The Cultural Selection of Animal Coat-Colour Diversity among the Bodi." In *Redefining Nature: Ecology, Culture and Domestication*, ed. R. Ellen and K. Fukui (pp. 319–386). Oxford: Berg.

Galaty, J. G.
1989 "Cattle and Cognition: Aspects of Maasai Practical Reasoning." In *The Walking Larder: Patterns of Domestication, Pastoralism and Predation*, ed. J. Clutton-Brock (pp. 215–230). London: Unwin Hyman.

Galbadrakh, D., and J. Ganchimeg.
2000 *Identification Guide of Common Plants of Mongolia*. Ulaanbaatar: Mongolian Society for Environmental Education.

Game, A.
2001 "Riding: Embodying the Centaur." *Body and Society* 7(1): 1–12.

Gauthier-Pilters, H., and A. I. Dagg.
1981 *The Camel: Its Evolution, Ecology, Behaviour and Relationship to Man*. Chicago: University of Chicago Press.

Gladitz, C.
1997 *Horse Breeding in the Medieval World.* Cornwall: Four Courts Press.

Goldstein, M. C., and C. M. Beall.
1994 *The Changing World of Mongolia's Nomads.* Berkeley: University of California Press.

Goodwin, D.
2002. "Horse Behaviour in Evolution, Domestication and Feralization." In *The Welfare of Horses,* ed. N. Waran (pp. 1–18). Dordrecht: Kluwer Academic.

Gould, S. J.
1979 "A Quahog is a Quahog." *Natural History* 88(7): 18–26.

Groves, C.
1994 "Morphology, Habitat, and Taxonomy." In *Przewalski's Horse: The History and Biology of an Endangered Species,* ed. L. Boyd and K. A. Houpt (pp. 39–60). New York: State University of New York Press.

Guilford, T., and M. S. Dawkins.
1991 "Receiver Psychology and the Evolution of Animal Signals." *Animal Behaviour* 42: 1–4.

Hallowell, I.
1976 "Ojibwa Ontology, Behavior, and World View." In *Contributions to Anthropology* (pp. 357–390). Chicago: University of Chicago Press.

Haraway, D.
2003 *The Companion Species Manifesto: Dogs, People and Significant Others.* Chicago: University of Chicago Press.
2004 *The Haraway Reader.* New York: Routledge.
2007 *When Species Meet (Posthumanities).* Minneapolis: University of Minnesota Press.

Hardin, G.
1971 "The Tragedy of the Commons." In *What a Piece of Work is Man: Introductory Readings in Biology,* ed. J. D. J. Ray and G. E. Nelson (pp. 351–364). Boston: Little, Brown.

Hare, B., M. Brown, C. Williamson, and M. Tomasello.
2002 "The Domestication of Social Cognition in Dogs." *Science* 298: 1634–1636.

Harris, D. R.
1996 "Domesticatory Relationships of People, Plants and Animals." In *Redefining Nature: Ecology, Culture and Domestication,* ed. R. Ellen and K. Fukui (pp. 437–465). Oxford: Berg.

Harris, M.
1992 "The Cultural Ecology of India's Sacred Cattle." *Current Anthropology* 33(1): 261–276.

Hart, B. L.
1985 *The Behaviour of Domestic Animals*. New York: Freeman.
Harvey, G.
2006a *Animism: Respecting the Living World*. New York: Columbia University Press.
2006b "Animals, Animists and Academics." *Zygon* 41(1): 9–19.
Haudricourt, A. G.
1969 "Domestication of Animals, Cultivation of Plants and Human Relations." *Social Science Information* 8(3): 163–172.
Hearne, V.
2000 *Adam's Task: Calling Animals by Name*. New York: Akadine Press.
Hemmer, H.
1990 *Domestication: The Decline of Environmental Appreciation*. Translated by N. Beckhaus. Cambridge: Cambridge University Press.
Heptner, V. G., A. A. Nasimovich, and A. G. Bannikov.
1961 *Mammals of the Soviet Union*, Vol. 1. Translated by P. M. Rao. Moscow: Vysshaya Shkola.
Heston, A.
1971 "An Approach to the Sacred Cow of India." *Current Anthropology* 12(2): 191–209.
Hilzheimer, M.
1935 "The Evolution of the Domestic Horse." *Antiquity* 9(34): 133–139.
Hooper, A.
1989 "Mutualism between Man and Honeyguide." In *The Walking Larder: Patterns of Domestication, Pastoralism and Predation*, ed. J. Clutton-Brock (pp. 347–349). London: Unwin Hyman.
Hovens, J. P. M., and K. Tungalaktuja.
2005 "Seasonal Fluctuations of the Wolf Diet in the Hustai National Park (Mongolia)." *Mammalian Biology* 4: 210–217.
Humphrey, C.
1980 "Theories of North Asian Shamanism." In *Soviet and Western Anthropology*, ed. E. Gellner (pp. 243–256). London: Gerald Duckworth.
1983 *The Karl Marx Collective: Economy, Society and Religion in a Siberian Collective Farm*. Cambridge: Cambridge University Press.
Humphrey, C., and D. Sneath.
1996 *Culture and Environment in Inner Asia, Vol. 1: The Pastoral Economy and the Environment*. Cambridge: Whitehorse Press.
1996 *Culture and Environment in Inner Asia, Vol. 2: Society and Culture*. Cambridge: Whitehorse Press.
1999 *The End of Nomadism? Society, State and the Environment in Inner Asia*. Durham, NC: Duke University Press.

Humphrey, C., with U. Onon.
1996 *Shamans and Elders: Experience, Knowledge and Power among the Daur Mongols*. Oxford: Clarendon Press.

Indra, R.
2003 "Mongolian Dairy Products." In *Mongolia Today: Science, Culture, Environment and Development*, ed. D. Badarch, R. A. Zilinskas, and P. J. Balint (pp. 69–85). London: RoutledgeCurzon.

Ingold, T.
1974 "On Reindeer and Men." *Man* 9(4): 523–538.
1980 *Hunters, Pastoralists and Ranchers: Reindeer Economies and Their Transformations*. Cambridge: Cambridge University Press.
1986a *The Appropriation of Nature: Essays on Human Ecology and Social Relations*. Iowa City: University of Iowa Press.
1986b "Reindeer Economies and the Origins of Pastoralism." *Anthropology Today* 2(4): 5–10.
1990 "An Anthropologist Looks at Biology." *Man* 25(2): 208–229.
1993 "The Temporality of the Landscape." *World Archaeology* 25: 152–174.
1994 "From Trust to Domination: An Alternative History of Human–Animal Relations." In *Animals and Human Society: Changing Perspectives*, ed. A. Manning and J. Serpell (pp. 1–22). New York: Routledge.
1996 "Hunting and Gathering as Ways of Perceiving the Environment." In *Redefining Nature: Ecology, Culture and Domestication*, ed. R. Ellen and K. Fukui (pp. 117–156). Oxford: Berg.
2000 *The Perception of the Environment: Essays on Livelihood, Dwelling and Skill*. New York: Routledge.

Jackson, M.
1983 "Knowledge of the Body." *Man* 18(2): 327–345.
1995 *At Home in the World*. Durham, NC: Duke University Press.

Jackson, R. R.
1992 "Eight-Legged Tricksters: Spiders that Specialize at Catching other Spiders." *Biosciences* 42: 590–598.

Jagchid, S.
2003 *Tibetan Buddhism, the Mongolian Religion*. Available at http://www.innermongolia.org/english/tibetan_buddhism.htm (accessed April 12, 2004).

Jansen, T., P. Forster, M. A. Levine, H. Oelke, M. Hurles, C. Renfrew, J. Weber, and K. Olek.
2002 "Mitochondrial DNA and the Origins of the Domestic Horse." *PNAS* 99(16): 10905–10910.

Jarman, M. R., G. Clark, C. Grigson, H.-P. Uerpmann and M. L. Ryder.
1976 "Early Animal Husbandry." *Philosophical Transactions of the Royal Society of London. Series B, Biological Sciences* 275(936): 85–97.

Jenkins, J. L.
1960 "The Morienhur: A Mongolian Fiddle." *Man* 60: 129–130.

Jewell, P A., C. Milner, and J. M. Boyd.
1974 *Island Survivors: The Ecology of the Soay Sheep of St. Kilda.* London: Athlone Press.

Jianlin, H., C. Richard, O. Hanotte, C. McVeigh, and J. E. O. Rege.
2002 *Yak Production in Central Asian Highlands.* Papers presented at the Proceedings of the Third International Congress on Yak, Lhasa, P. R. China. International Livestock Research Institute, Nairobi.

Jila, N.
2006 "Myths and Traditional Beliefs about the Wolf and the Crow in Central Asia." *Asian Folklore Studies* 65: 161–177.

Johnson, D. A., D. P. Sheehy, D. Miller, and D. Damiran.
2006 "Mongolian Rangelands in Transition." *Sécheresse* 17(1–2): 133–141.

Kaczensky, P.
2007 "Wildlife Value Orientations of Rural Mongolians." *Human Dimensions of Wildlife* 12: 317–329.

Kaminski, J., J. Riedel, J. Call, and M. Tomasello.
2005 "Domestic Goats, *Capra hircus*, Follow Gaze Direction and Use Social Cues in an Object Choice Task." *Animal Behaviour* 69: 11–18.

Khazanov, A. M.
1994 *Nomads and the Outside World,* 2nd ed. Translated by J. Crookenden. Madison: University of Wisconsin Press.

King, A. D.
2003 "Without Deer There is no Culture, Nothing." *Anthropology and Humanism* 27(2): 133–164.

King, S. R. B., and J. Gurnall.
2005 "Habitat Use and Spatial Dynamics of Takhi Introduced to Hustai National Park, Mongolia." *Biological Conservation* 124: 277–290.
2007 "Scent-Marking Behaviour by Stallions: An Assessment of Function in a Reintroduced Population of Przewalski Horses (*Equus ferus przewalskii*)." *Journal of Zoology* 272: 30–36.

Knight, J., ed.
2005 *Animals in Person.* Oxford: Berg, Oxford.

Kohn, E.
2007 "How Dogs Dream: Amazonian Natures and the Politics of Transpecies Engagement." *American Ethnologist* 34(1): 3–24.

Latour, B.
1993 *We Have Never Been Modern.* Cambridge, MA: Harvard University Press.

Lattimore, O.

1962 *Nomads and Commissars: Mongolia Revisited.* New York: Oxford University Press.

Laufer, B.

1930 "The Early History of Felt." *American Anthropologist* 32(1): 1–18.

Lawrence, E. A.

1990 Rodeo Horses: The Wild and the Tame. In *Signifying Animals: Human Meaning in the Natural World,* ed. R. G. Willis (pp. 178–195). One World Archaeology Series. London: Unwin Hyman.

Leeds, A., and A. P. Vayda.

1965 *Man, Culture and Animals: The Role of Animals in Human Ecological Adjustments,* No. 78. Washington, DC: American Association for the Advancement of Science.

Lestel, D.

1998. "Chimpanzees Have Domesticated Humans: Towards an Anthropology of Human–Animal Communication." *Anthropology Today* 14(3): 12–15.

2006 "Ethology and Ethnology: The Coming Synthesis. A General Introduction." *Social Science Information* 45(2): 147–153.

Lestel, D, F. Brunois, and G. Gaunet.

2006 "Etho-Ethnology and Ethno-Ethology." *Social Science Information* 45(2): 155–177.

Levine, M.

2004 "Exploring the Criteria for Early Horse Domestication." In *Traces of Ancestry: Studies in Honour of Colin Renfrew,* ed. M. Jones (pp. 115–126). McDonald Institute Monograph, McDonald Institute for Archaeological Research, Cambridge.

Levine, M. A., G. N. Bailey, K. E. Whitwell, and L. B. Jeffcott.

2000 "Palaeopathology and Horse Domestication: The Case of Some Iron Age Horses from the Altai Mountains, Siberia." In *Human Ecodynamics and Environmental Archaeology,* ed. G. N. Bailey, R. Charles, and N. Winder (pp. 123–133). Oxford: Oxbow.

Ligaa, U., and D. Tsembal.

2003 "Medicinal Plants of Mongolia and Their Use in Traditional Medicine." In *Mongolia Today: Science, Culture, Environment and Development,* ed. D. Badarch, R. A. Zilinskas, and P. J. Balint (pp. 69–85). London: RoutledgeCurzon.

Linklater, W. L.

2000 "Adaptive Explanation in Socio-Ecology: Lessons from the Equidae." *Biological Review* 75: 1–20.

Lopez, B.

1978 *Of Wolves and Men.* New York: Scribner's.

Lott, D. F., and B. L. Hart.
1977 "Aggressive Domination of Cattle by Fulani Herdsmen and Its Relation to Aggression in Fulani Culture and Personality." *Ethos* 5(2): 174–186.
1979 "Applied Ethology in a Nomadic Cattle Culture." *Applied Animal Ethology* 5: 309–319.

Lusis, J. A.
1941. "Striping Patterns in Domestic Horses." *Genetica*, 23(1): 31–62.

MacDougall, D.
1992. "Complicities of Style." In *Film as Ethnography*, ed. P. I. Crawford and D. Turton (pp. 90–98). Manchester: Manchester University Press.
1998 *Transcultural Cinema*. Princeton, NJ: Princeton University Press.
2006 *The Corporeal Image: Film, Ethnography and the Senses*. Princeton, NJ: Princeton University Press.

Macuda, T., and B. Timney.
1999. "Luminance and Chromatic Discrimination in the Horse (*Equus caballus*)." *Behavioural Processes* 44(3): 301–307.

Magash, A.
2003 *Yak in Mongolia*. Bangkok, Thailand: Food and Agriculture Organization of the United Nations, Regional Office for Asia and the Pacific.

Martin, L.
1986 "'Eskimo Words for Snow': A Case Study in the Genesis and Decay of an Anthropological Example." *American Anthropologist* 88(2): 418–423.

Marx, K.
1976 *Capital*, 4th German ed. Translated by E. A. C. Paul. London: Dent. (Original work published 1867)

Mauss, M.
1985 "A Category of the Human Mind: The Notion of Self." In *The Category of the Perso: Anthropology, Philosophy, History*, ed. M. Carrithers, S. Collins, and S. Lukes (pp. 1–25). Cambridge: Cambridge University Press. (Original work published 1938)

Mayhew, B.
2001 *Lonely Planet Mongolia*, 3rd ed. Victoria, Australia: Lonely Planet Publications.

Mayor, A.
1991 "Griffin Bones: Ancient Folklore and Paleontology." *Cryptozoology* 10: 16–41.

McDermott, J. P.
1989 "Animals and Humans in Early Buddhism." *Indo-Iranian Journal* 32: 269–280.

McKinley, J., and T. D. Sambrook.
2000 "Use of Human-Given Cues by Domestic Dogs (*Canis familiaris*) and Horses (*Equus caballus*)." *Animal Cognition* 3: 13–22.

Midgley, M.
1985 "Persons and Non-Persons." In *In Defence of Animals*, ed. P. Singer (pp. 52–62). New York: Basil Blackwell.

Miller, D. J.
1999 "Nomads of the Tibetan Plateau Rangelands in Western China, Part Two: Pastoral Production Practices." *Rangelands* 21(1): 16–19.

Mirov, N. T.
1945 "Notes on the Domestication of Reindeer." *American Anthropologist* 47(3): 393–408.

Miyazaki, S., T. Yasunari, and T. Adyasuren.
1999 "Abrupt Seasonal Changes of Surface Climate Observed in Northern Mongolia by an Automatic Weather Station." *Journal of the Meteorological Society of Japan* 77(2): 583–593.

Mullin, M. H.
1999 "Mirrors and Windows: Sociocultural Studies of Human–Animal Relationships." *Annual Review of Anthropology* 28: 201–224.

Nelson, R. K.
1983 *Make Prayers to the Raven: A Koyukon View of the Northern Forest.* Chicago: University of Chicago Press.

Neupert, R. F.
1999 "Population, Nomadic Pastoralism and the Environment in the Mongolian Plateau." *Population and the Environment* 20(5): 29.

Newby, J.
1997 *The Pact for Survival: Humans and Their Companion Animals.* Sydney: ABC Books.

Nickel, H.
1977 "And Behold, A White Horse: Observations on the Colours of the Horses of the Four horsemen of the Apocalypse." *Metropolitan Museum Journal* 12: 179–183.

Noske, B.
1989 *Humans and Other Animals: Beyond the Boundaries of Anthropology.* London: Pluto Press.
1993 "The Animal Question in Anthropology." *Society and Animals* 1(2): 185–190.

O'Connor, T. P.
1996 "Working at Relationships: Another Look at Animal Domestication." *Antiquity* 71: 149–156.

Olupona, J. K.
1993 "Some Notes on Animal Symbolism in African Religion and Culture." *Anthropology and Humanism* 18(1): 3–12.

Open Society Forum.
2004 *The Future of Nomadic Pastoralism in Mongolia: Public Perception Survey.* Ulaanbaatar: Open Society Forum.

Onon, U.
2001 *The Secret History of the Mongols: The Life and Times of Chinggis Khan.* Richmond, Surrey: Curzon Press.

Ortner, S.
1973 "On Key Symbols." *American Anthropologist* 75(5): 1338–1346.

Osmaston, H., R. Fisher, J. Frazer, and T. Wilkinson.
1994 "Animal Husbandry in Zangskar." In *Himalayan Buddhist Villages*, ed. J. Crook and H. Osmaston (pp. 199–248). Bristol: University of Bristol.

Patton, P.
2003 "Language, Power and the Training of Horses." In *Zoontologies: The Question of the Animal*, ed. C. Wolfe (pp. 83–99). Minneapolis: University of Minnesota Press.

Pederson, M. A.
2001 "Totemism, Animism and North Asian Indigenous Ontologies." *The Journal of the Royal Anthropological Institute* 7(3): 411–427.

Pennisi, E.
2001 "Horses Domesticated Multiple Times." *Science* 291(5503): 412.

Phillips, M. T.
1994 "Proper Names and the Social Construction of Biography: The Negative Case of Laboratory Animals." *Qualitative Sociology* 17(2): 119–142.

Phillips, R. W., I. A. Tolstoy, and R. G. Johnson.
1946 "Yaks and Yak-Cattle Hybrids in Asia." *Journal of Heredity* 37: 162–170.

Plumwood, V.
2003 *Animals and Ecology: Towards a Better Integration.* Available at http://socpol.anu.edu.au/pdf-files/Vegpap6%20%20.pdf (accessed March 26, 2008).

Potapov, L. P.
1999 "Shaman's Drum: A Unique Monument of Spiritual Culture of the Altai Turk Peoples." *Anthropology of Consciousness* 10(4): 24–35.

Pratt, D. G., D. C. MacMillan, and I. J. Gordon.
2004 "Local Community Attitudes to Wildlife Utilisation in the Changing Economic and Social Context of Mongolia." *Biodiversity and Conservation* 13: 591–613.

Price, E. O.
1999 "Behavioral Development in Animals Undergoing Domestication." *Applied Animal Behaviour Science* 65: 245–271.

Rabey, M. A.
1989 Are Llama-Herders in the South Central Andes True Pastoralists? In *The Walking Larder: Patterns of Domestication, Pastoralism and Predation*, ed. J. Clutton-Brock (pp. 269–176). London: Unwin Hyman.

Rindos, D.
1980 "Symbiosis, Instability, and the Origins and Spread of Agriculture: A New Model." *Current Anthropology* 21(6): 751–772.

Rose, D. B.
1984. *Dingo Makes Us Human: Being and Purpose in Aboriginal Culture.* Cambridge: Cambridge University Press.

Roux, J.-P.
2003. *Ghenghis Khan and the Mongol Empire.* Translated by T. Ballas. Thames: New Horizons.

Rudenko, S. I.
1970 *Frozen Tombs of Siberia: The Pazyryk Burials of Iron Age Horsemen.* London: Dent.

Samashev, Z., and H.-P. Francfort.
2002 "Scythian Steeds." *Archaeology* May/June: 32–35.

Sanders, A. J. K., and J. Bat-Ireedui.
1999 *Colloquial Mongolian: The Complete Course for Beginners.* New York: Routledge.

Savolainen, P., Y. Zhang, J. Luo, J. Lundeberg, and T. Leitner.
2002 "Genetic Evidence for an East Asian Origin of Domestic Dogs." *Science* 298: 1610–1613.

Schaller, G. B.
1998 *Wildlife of the Tibetan Steppe.* Chicago: University of Chicago Press.

Scott, C.
1996 *"Science for the West, Myth for the Rest?"* In *Naked Science: Anthropological Inquiry into Boundaries, Power and Knowledge*, ed. L. Nader (pp. 71–86). New York: Routledge.

Serpell, J. A.
1989 "Pet-Keeping and Animal Domestication: A Reappraisal." In *The Walking Larder: Patterns of Domestication, Pastoralism and Predation*, ed. J. Clutton-Brock (pp. 10–19). London: Unwin Hyman.
1995 *The Domestic Dog: Its Evolution, Behaviour and Interactions with People.* Cambridge: Cambridge University Press.
2000 "Creatures of the Unconscious: Companion Animals as Mediators." In *Companion Animals and Us: Exploring the Relationships Between People and Pets*, ed. A. L. Podberscek, E. S. Paul, and J. A. Serpell (pp. 108–121). Cambridge: Cambridge University Press.
2005 "Animals and Religion: Towards a Unifying Theory." In *The Human–Animal Relationship*, ed. F. N. De Jong and R. Van den Bos (pp. 9–22). Assen, Netherlands: Royal Van Gorcum.

Shanklin, E.
1985 "Sustenance and Symbol: Anthropological Studies of Domesticated Animals." *Annual Review of Anthropology* 14: 375–403.

Siklos, B.
1991 "Mongolian Buddhism: A Defensive Account." In *Mongolia Today*, ed. S. Akiner (pp. 155–182). London: Kegan Paul.

Simoons, F. J.
1968 *A Ceremonial Ox of India: The Mithan in Nature, Culture, and History, with Notes on the Domestication of Common Cattle*. Madison: University of Wisconsin Press.

Simpson, G. G.
1938 "Mongolian Mammal Names." *American Museum Novitates* 980: 1–26.

Smuts, B.
2001 "Encounters with Animal Minds." *Journal of Consciousness Studies* 8(5–7): 293–309.

Sneath, D.
2000 *Changing Inner Mongolia: Pastoral Mongolian Society and the Chinese State*. Oxford: Oxford University Press.

Sorokin, S.
1990 "Horse-Drawn Vehicles of the Eurasian Forest-Steppe in 'Pre-Centaurian' Times." *Iranica Antiqua* 25: 97–147.

Stewart, S.
2001 *In the Empire of Genghis Khan: A Journey among Nomads*. London: HarperCollins.

Strathern, M.
1980 "No Nature, No Culture: The Hagen Case." In *Nature, Culture and Gender*, ed. C. MacCormack and M. Strathern (pp. 174–222). Cambridge: Cambridge University Press.

Szynkiewicz, S.
1990 "Sheep Bone as a Sign of Human Descent: Tibial Symbolism among the Mongols." In *Signifying Animals: Human Meaning in the Natural World*, ed. R. G. Willis (pp. 74–84). One World Archaeology Series. London: Unwin Hyman Ltd.

Tambiah, S. J.
1969 "Animals are Good to Think and Good to Prohibit." *Ethnology* 8(4): 423–452.

Tani, Y.
1989 "The Geographical Distribution and Function of the Sheep Flock Leader." In *The Walking Larder: Patterns of Domestication, Pastoralism and Predation*, ed. J. Clutton-Brock (pp. 185–198). London: Unwin Hyman.
1996. "Domestic Animal as Serf: Ideologies of Nature in the Mediterranean and the Middle East." In *Redefining Nature: Ecology,*

Culture and Domestication, ed. R. Ellen and K. Fukui (pp. 387–416). Oxford: Berg.

Telenged, B.
1996 "Livestock Breeding in Mongolia Past and Present: The Advantages and Disadvantages of Traditional and Modern Animal Breeding Practices." In *Culture and Environment in Inner Asia, Vol. 1: The Pastoral Economy and the Environment*, ed. C. Humphrey and D. Sneath (pp. 161–184). Cambridge: Whitehorse Press.

Tervish, L.
2006 *Mongol Zurkhain Tsag Toonii Bichig 2007–2008* [Esteemed Writing of the Mongolian Astrological Calendar]. Ulaanbaatar: Publisher unknown.

Tomasello, M., and J. Call.
2004 "The Role of Humans in the Cognitive Development of Chimpanzees Revisited." *Animal Cognition* 7: 213–215.

Tomikawa, R.
2006 "Mongolian Wrestling (*Bukh*) and Ethnicity." *International Journal of Sport and Health Sciences* 4: 103–109.

Trut, L. N.
1999 "Early Canid Domestication: The Farm Fox Experiment." *American Scientist* 87(2): 11.

Tseren, P. B.
1996 "Traditional Pastoral Practice of the Oirat Mongols and Their Relationship with the Environment." In *Culture and Environment in Inner Asia, Vol. 1: The Pastoral Economy and the Environment*, ed. C. Humphrey and D. Sneath (pp. 147–159). Cambridge: Whitehorse Press.

Tsing, A. L.
2004 *Friction: An Ethnography of Global Connection*. Princeton, NJ: Princeton University Press.
In press "Unruly Edges: Mushrooms as Companion Species." In *Thinking with Donna Haraway*, ed. S. Ghamari-Tabrizi, Cambridge, MA: MIT Press.

Tumerjav, M.
2003 "Traditional Animal Husbandry Techniques Practiced by Mongolian Nomadic People." In *Mongolia Today: Science, Culture, Environment and Development*, ed. D. Badarch, R. A. Zilinskas, and P. J. Balint (pp. 86–113). London: RoutledgeCurzon.

Uexküll, J von.
1957 "A Stroll Through the Worlds of Animals and Men: A Picture Book of Invisible Worlds." In *Instinctive Behaviour: The Development of a Modern Concept*, ed. C. H S. Schiller (pp. 5–80). New York: International University Press.

Vainshtein, S. I.
1980 *Nomads of South Siberia: The Pastoral Economies of Tuva*. Cambridge: Cambridge University Press. (Original work published 1972)

Vila, C., J. A. Leonard, A. Gotherstrom, S. Marklund, K. Sandberg, K. Liden, R. K. Wayne, and H. Ellegren.
2001 "Widespread Origins of Domestic Horse Lineages." *Science* 291: 474–477.

Vitebsky, P.
2005 *Reindeer People: Living with Animals and Spirits in Siberia*. London: HarperCollins.

Viveiros de Castro, E.
1998 "Cosmological Deixis and Amerindian Perspectivism." *The Journal of the Royal Anthropological Institute* 4(3): 469–488.

Vreeland, H. H.
1957 *Mongol Community and Kinship Structure*. Behaviour Science Monographs, Human Relations Area Files, New Haven, CT.

Waddington-Humphrey, C.
1974 "Horse Brands of the Mongolians: A System of Signs in a Nomadic Culture." *American Ethnologist* 1(3): 471–488.

Walther, F. R.
1991 "On Herding Behavior." *Applied Animal Behaviour Science* 29: 5–13.

Yembuu, B., and Munkh-Erdene, K.
2005 *Literacy Country Study: Mongolia*. Ulaanbaatar, Mongolia: UNESCO.

Zeder, M. A., E. Emshwiller, B. D. Smith, and D. G. Bradley.
2006 "Documenting Domestication: The Intersection of Genetics and Archaeology." *TRENDS in Genetics* 22(3): 139–155.

Zeuner, F. E.
1963 *A History of Domesticated Animals*. London: Hutchinson.

Zhang, R. C.
2000 "Ecology and Biology of Yak Living in Qinghai-Tibetan Plateau." In *Recent Advances in Yak Reproduction*, ed. X. X. Zhao and R. C. Zhang (pp. 1–5). New York: International Veterinary Information Service.

INDEX

Printed in Great Britain
by Amazon

14541030R00173